A NON-MATHEMATICAL
APPROACH TO BASIC MRI

A NON-MATHEMATICAL APPROACH TO BASIC MRI

Hans-Jørgen Smith, M.D.
Head of MRI
Department of Radiology
Rikshospitalet, Oslo

Frank N. Ranallo, M.S.
Director of Radiological Physics Services
Department of Medical Physics
University of Wisconsin-Madison

1989

Medical Physics Publishing Corporation
Madison, Wisconsin

Published by:

Medical Physics Publishing Corporation
1300 University Avenue
Madison, Wisconsin 53706

ISBN: 0-944838-02-2

Library of Congress Cataloging in Publication Data:

Smith, Hans-J. (Hans-Jørgen), 1948-
    A non-mathematical approach to basic MRI.

    Includes bibliographies and index.
    1. Magnetic resonance imaging. 2. Nuclear magnetic resonance.
I. Title. [DNLM: 1. Nuclear Magnetic Resonance. QC 762 S649n]
RC78.7.N83S65   1989   616.07'57   89-28170

We are indebted to GE Medical Systems for the use of the photograph of the MR imager which appears on the cover and on page 3 of the text.

Cover design by Becky Chapman-Winter

to

Tone Lise and Veronika

H.-J. S.

to

Gail

F.N.R.

Table of Contents

Section I. Basic Principles of NMR

Section II. Relaxation & Image Contrast

Section III. MR Imaging Methods

Foreword

Magnetic Resonance Imaging (MRI) was introduced into clinical practice in 1981 and has undergone unprecedented growth and development. Current MR scanners are capable of producing images with unique contrast and superb spatial resolution. Accompanying the development of clinical MR imaging has been a parallel increase in the complexity of the MR scanning techniques. It is therefore more important than ever for the radiologist or medical specialist working with MRI to have a clear understanding of the basic principles involved in generating the MR image.

Many attempts have been made to explain MR phenomenon to the radiological community. This textbook stands apart from other discussions by virtue of its clarity and accuracy. The material is presented in a fashion that can be easily understood by the clinical radiologist. The discussions of image acquisition and image contrast are superb overviews of these topics. The authors have included sections on chemical shift imaging and MR flow effects, bringing the reader up to date on recent advances in MR imaging. The chapters on contrast enhanced MR imaging and gradient echo imaging are particularly helpful and provide a basis for understanding many of the observations that are encountered clinically. The book's emphasis on a non-mathematical approach is also a welcome change from the complex discussions that are frequently encountered in the MR literature.

I commend the authors for succeeding in producing an understandable and readable textbook for individuals new to the field of MR imaging. The concepts presented in this book provide the foundation that is necessary for understanding many of MR imaging techniques that are commonly used today.

Patrick Turski, M.D.
Associate Professor of Radiology and
Chief of Magnetic Resonance Imaging Section,
Department of Radiology,
University of Wisconsin

Magnetic resonance imaging (MRI) is still a new diagnostic modality. The physical principles of MRI are rather complex and have little in common with those of the more established radiological methods using x-rays, radioisotopes, or ultrasound. Learning about MRI is therefore a challenge for technologists, residents, and radiologists. Usually they do not have the background in mathematics and physics needed to understand the numerous articles or textbooks in physics dealing with basic MRI. The mere sight of an exponential formula can, to many, be quite discouraging. On the other hand, the more superficial, non-mathematical approaches to MRI usually leave many questions unanswered, and may even add to the reader's confusion by using inappropriate analogies.

This book is an attempt to fill the gap between the physicists' world of complicated mathematical formulas and popularized versions of MRI basics. This is not meant to be a complete textbook on the physical principles of MRI; technical aspects have largely been omitted. Its purpose is to provide the reader with the conceptual foundations of magnetic resonance and magnetic resonance imaging; and to help the reader understand the roles of the major parameters involved in the making of an MR image, and the consequences of changing them.

This book should enable the reader to understand conventional techniques such as partial saturation, spin-echo and inversion-recovery as well as the newer fast imaging techniques with small flip angles and gradient echoes. The fundamentals of special topics such as contrast enhanced imaging, flow effects, and chemical shift imaging are also covered.

The use of mathematics in this book is kept to a bare minimum of essential formulas. For many readers these formulas will assist in their understanding of MRI phenomena. As each formula is introduced, however, its meaning will be explained in the text so that the less mathematically inclined reader will not be confused. Further mathematical material, of interest to some readers, is set off in sections labeled "A Mathematical Interlude." These sections can be omitted by the reader, if desired, without harming the continuity of the text. In all cases, the mathematics is restricted to the level of basic algebra.

Acknowledgements

This book could not have become a reality without the support from the Department of Radiology, University Hospital and Clinics, Madison, Wisconsin. I especially want to thank Professor and Chairman Joseph F. Sackett for giving me the opportunity to work at his department, Drs. Charles M. Strother and Patrick A. Turski for encouraging me to write this book, Ms. Sandy Yost for typing the manuscript and the Department of Medical Illustration, University of Wisconsin, Madison, for making the illustrations.

Hans-J. Smith

I would like to express my graditude to several individuals who provided assistance and encouragement in the preparation of this book. Tom Foo, Dr. Curtis Partington, Dr. Siamak Shahabi, and Dr. Patrick Turski are among the members of the U.W. Departments of Medical Physics and Radiology whose discussions concerning MRI processes I greatly appreciate. The transformation of sketches and ideas into the illustrations in this book are due to the excellent, and much appreciated, efforts of Betsy True of the U.W. Department of Medical Illustrations. I am indebted to the UW Department of Radiology for funding her work of illustrating this book. I would like to thank Dr. John Cameron, Professor and Chairman Emeritus of the University of Wisconsin Department of Medical Physics for his ideas, editorial assistance, and patient encouragement. My thanks also go to Professor P.R. Moran from whom I first learned the correct ways of analyzing nuclear magnetic resonance and digital imaging phenomena. Finally I would like to express my sincere appreciation to my wife for her understanding and for her help in proofreading this manuscript.

Frank Ranallo

1. Introduction

Nuclear magnetic resonance (NMR) as a physical phenomenon was first described by Block and Purcell in 1946, an achievement for which they both received the Nobel prize in physics in 1951. The earliest application of the phenomenon was NMR spectroscopy, which revealed detailed information of the structure of molecules and the molecular composition of chemical substances. NMR spectroscopy is still a very important analytical method in physics and chemistry, but the spectrum obtained does not contain any spatially dependent information about chemical processes occurring at differing locations. By the early fifties, researchers realized that NMR signals could be used to obtain spatial information, but a serious attempt to actually construct an NMR imager did not take place until 1973 after Lauterbur described his imaging technique called zeugmatography. In 1975 EMI began development of a commercial magnetic resonance imaging (MRI) system, and from then on, progress in this technology has been rapid.

The MRI systems of today can provide images in any plane through any part of the body. The acquisition time for an image has been reduced from approximately one hour to a few minutes or less, and the best spatial resolution is less than 1 mm. A modern MRI unit is shown in Fig. 1-1.

Computed tomography (CT) and MRI share some significant similarities. Both are tomographic modalities that produce images of slices through a selected part of the body. Fig. 1-2 shows images of similar parts of the brain produced by CT and by MRI systems. Superficially, these two images and even the outward appearance of the equipment that produced them — their control consoles and gantry assemblies — appear similar; however, CT and MRI are different in some very important ways.

CT uses x-rays to produce images. The CT image is derived from measurements of the attenuation of x-ray beams passing through the slice being imaged. Contrast in a CT image is provided by differences in the attenuation of these x-ray beams and is thus dependent on only two parameters: electron density and effective atomic number of the tissues.

To produce images, MRI uses magnetic fields and radio waves which are believed to pose less risk to the patient than x-rays. A magnetic resonance image is derived from radio wave signals coming from the hydrogen nuclei (protons) in the body soon after they have absorbed energy from radio wave signals transmitted into the body.* The contrast in an MRI image is provided by differences in the signal intensity from these nuclei in the various tissues. The differences in signal intensity are determined by at least 4 different parameters: the density (concentration) of the protons, the "behavior" of the protons as determined by two different time constants, (T1 and T2), and finally, by the bulk flow of protons, whether in blood or CSF.

This difference in the source of image contrast between CT and MRI is of crucial importance. X-ray attenuation is a physical property of tissue that is unaffected by its "chemistry"; therefore, x-ray attenuation is determined by the types and amounts of chemical elements present but not by their motions or chemical interactions. MRI signals from the hydrogen nuclei, however, are strongly affected by the chemical interactions, molecular motions, and fluid flow occurring in the tissue and thus are sensitive to an entire spectrum of phenomena to which CT is totally blind.

It may be helpful here to take a quick look at the various steps leading to the creation of the MR signal. When a patient is studied with MRI, he or she is placed in the bore of a large magnet. Through that bore there is a very strong, static magnetic field, which in most MRI machines is oriented along the long axis of the patient's body. Each of the hydrogen nuclei, or protons, in the body can be looked upon as a tiny, spinning magnet whose north-south magnetic axis is the same as its spin axis. When placed in a strong, static magnetic field, the axes of these protons will tend to align along the direction of the magnetic field.

To initiate the imaging process, a short pulse of radio frequency electromagnetic energy (radio waves) is transmitted into the body. The frequency and intensity of this pulse can be adjusted so that some of its energy will be absorbed by the protons and cause their magnetic (or spin) axes to rotate, on average, by 90° to a plane perpendicular to the static magnetic field. Due to their spin, the protons' axes will wobble (or precess) around the direction of the static magnetic field, similar to how a spinning top on a table precesses around the direction of gravity (the vertical direction). This precession of the tiny magnets, the protons, induces an electric

---

* The nuclei of all elements contain protons; however, throughout this book we will frequently use the term "proton" to mean specifically the hydrogen nucleus, which is composed of a single proton, and the term "protons" to refer to a number of hydrogen nuclei.

current in receiver coils placed outside of the body, but inside the bore of the magnet. The induced current is the MR signal that can be transformed by a computer into an image.

The above is a very superficial look at the origin of the MR signal. To make it all understandable we have to go back and describe each step in more detail. We start by looking at some of the properties of atomic nuclei.

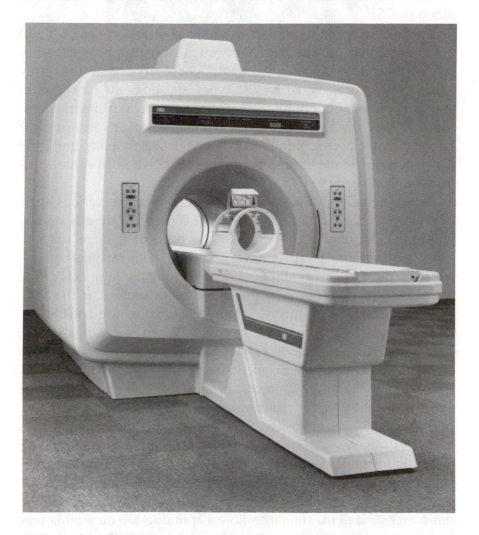

Fig. 1-1. A modern MR unit (courtesy of GE Medical Systems).

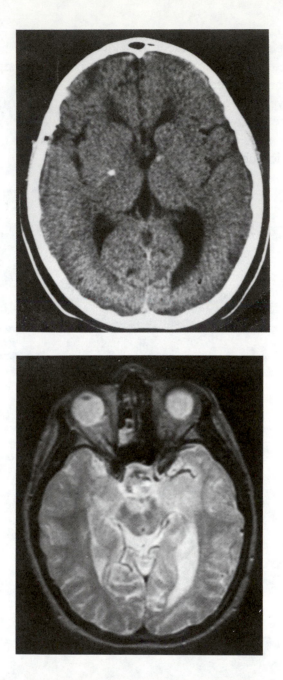

Fig. 1-2. A CT (top) and MR (bottom) image of an axial slice through the brain. Both images are of the same individual. The MR image shows a plane that passes through both eyes and the lower portion of the brain. This image demonstrates a mass behind the orbits and in front of the brain stem. The CT image shows a somewhat similar plane in the brain though at a different angle (less horizontal) so that the image plane passes above the position of the orbits.

I. Basic Principles of NMR

2. Nuclear Spin, Nuclear Magnetic Moment

All atomic nuclei are composed of protons and neutrons. Protons are minute particles of positive charge, while neutrons have about the same mass as protons but have no overall electric charge. The simplest nucleus is that of hydrogen, which consists of a single proton. Both protons and neutrons have a property called **spin** or **angular momentum**. Though this property belongs to the somewhat mysterious realm of quantum physics, in this book it will do no harm for us to visualize spin as an actual physical rotation similar to the earth's rotation on its axis or the spin of a top.

In addition to its spin, the proton also has a **magnetic moment**, which means that it behaves like a magnet. The reason why the proton can be looked upon as a tiny magnet is twofold: (1) the proton has an electric charge, and (2) it rotates around its own axis in the motion called spin. Any electrically charged object that is moving will surround itself with a magnetic field, and when the motion is that of a spin, the object is referred to as a **magnetic dipole**. A proton is thus a magnetic dipole (Fig. 2-1a). Its magnetic field will have a configuration similar to that of a small bar magnet with a north and south pole.

A magnetic dipole not only produces a magnetic field, but also responds to the presence of any magnetic field from other sources. In Chapters 3 through 6 we will be concerned principally with how a magnetic dipole reacts to externally applied, static and oscillating magnetic fields.* In Chapter 7 we will also be interested in how the proton's own magnetic field produces the MR signal.

At this point we need to discuss a fairly simple but important mathematical concept that we will frequently use in the rest of this book. Many types of measurements in science can be described by a simple number (with units), for example, temperature, voltage, and energy. However, some properties need to be described by both a number with units (giving the size, strength, or magnitude) *and* a direction. Such a specification, using both a number and a direction, is called a **vector**. The velocity of a car is a common example of a vector, since to completely describe a car's velocity

---

* By "static" we mean constant in time. Here, of course, we are referring to the magnetic field produced by the main magnet of the MRI scanner.

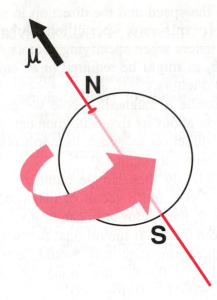

Fig. 2-1a. The proton viewed as a magnetic dipole. The positively charged proton spins and therefore behaves like a magnet with a north pole (N) and a south pole (S). The strength and orientation of this magnet is given by the magnetic moment vector, μ.

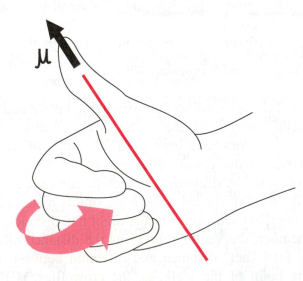

Fig. 2-1b. The proton's magnetic moment vector, μ, is oriented along the spin axis. The direction of the vector μ is determined by the direction of rotation of the proton's spin. If the fingers of the right hand are curled in the direction of the rotation, the thumb points in the general direction of the magnetic moment.

we must give both the speed and the direction in which the car is heading. In the text of this book we will follow the common practice of using **bold** letters when specifying vectors. To specify just the size of a vector, as might be required in an equation, we will use regular, non-bold letters.

To properly describe a magnetic dipole we need to give two pieces of information about its magnetic moment: its strength and its orientation (the direction in which its north pole is pointing). Thus magnetic moment is a vector.

Vectors add differently than simple numbers. While 2 + 2 is always equal to 4, if you add a vector of magnitude 2 to another vector of magnitude 2, the **vector sum** can be a vector with a magnitude anywhere from 0 to 4. If the original two vectors pointed in exactly the same direction, their sum would be a vector with a magnitude of 4, while if the vectors pointed in exactly opposite directions their sum would be 0 (Fig. 2-2).

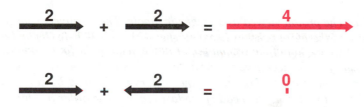

Fig. 2-2. An example of the addition of two vectors of equal size which point a) in the same direction, and b) in opposite directions.

Figure 2-3 shows a method of visualizing vector addition. When adding vectors the tail of one vector is placed at the head of the other until all vectors have been used. The order of adding vectors has no effect on the result. The sum of the vectors or the **net vector** is obtained by drawing a line connecting the tail of the first vector to the head of the last. The logic of this method can be seen by imagining that each vector is a straight line distance walked in a field (remember, the vector represents the distance *and* the direction). The total "net" distance and direction between the starting and ending point of the walk, as the crow flies, is then the net vector.

The strength and orientation of a magnetic dipole is given by a vector called the **magnetic dipole moment** or simply the magnetic moment. In Figure 2-1a the magnetic moment of the proton is represented by the arrow labeled μ. The length of this arrow indicates

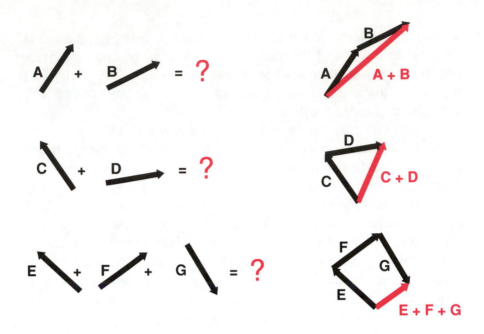

Fig. 2-3. The general method of vector addition. To add together 2 or more vectors, place the tail of the second vector at the head of the first; continue placing the tail of each successive vector at the head of the previous vector until all vectors have been used. The sum of the vectors is obtained by connecting the tail of the first vector to the head of the last vector. The order in which vectors are added has no effect on the result.

the strength of the magnetic moment; its direction indicates the orientation of magnetic moment. For a proton, the direction of the magnetic moment is oriented along the proton's spin axis. Spin is also a vector since it has a size and a direction. The spin vector points along the spin axis in the general direction indicated by the thumb if the fingers of the right hand are curled in the direction of the rotation (Fig. 2-1b). For a proton, the spin and magnetic moment point in the same direction.

Several protons considered together have a composite magnetic moment or **net magnetic moment** which is the vector sum of the magnetic moments of each of the protons. Figure 2-4 shows an example of how the magnetic moment vectors for a small number of protons can add together to yield a net magnetic moment. Generally, if the individual magnetic moments point in different directions, they will tend to cancel and result in a small net magnetic moment. If instead they point in the same general direction, the result will be a larger net magnetic moment.

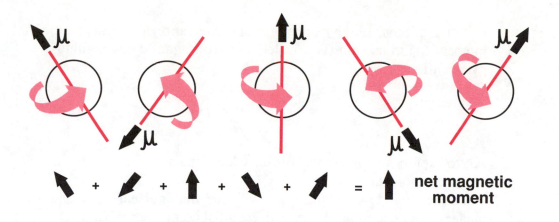

Fig. 2-4. *The net magnetic moment that results from the addition of the magnetic moments (μ) from 5 protons. This figure shows one specific group of orientations for the magnetic moments of the protons. Other possible orientations would yield net magnetic moments that are from 0 to 5 times the magnetic moment of a single proton. The net magnetic moments that we are concerned with in the body are, of course, due to much larger numbers of protons.*

Since the hydrogen nucleus consists of a single proton, it has both spin and magnetic moment. The hydrogen nucleus, or proton, is the most suitable nucleus for MRI because of its abundance in the body, and because the single proton produces the largest signal of all common stable atomic nuclei (at a constant magnetic field). The other atomic nuclei in the body consist of a various number of protons and neutrons and all of these protons and neutrons have spin. In the nucleus the proton spins form pairs whose spins point in opposite directions and therefore cancel out. Neutrons behave in a like fashion. Thus, if the number of protons in a nucleus is even, their spins will cancel out, and the same applies to an even number of neutrons. The nucleus as a whole will therefore have spin only when it contains an odd number of protons and/or an odd number of neutrons. No net nuclear spin means no nuclear magnetic moment, and without a magnetic moment, the nucleus cannot respond to a radio frequency pulse nor can it induce a current in a receiver coil. Thus it is useless for MRI. This is why oxygen — ^{16}O, with 8 protons and 8 neutrons — and the common isotope of carbon — ^{12}C, with 6 protons and 6 neutrons — which both are abundant in human tissue, cannot be used for MRI (or MR spectroscopy, for that matter). There are some useful nuclei besides hydrogen, such as ^{13}C, ^{23}Na, ^{19}F and ^{31}P, but the use of these for imaging is still experimental. In this book, only proton imaging will be dealt with.

For a proton, or any other nucleus, the ratio of its magnetic moment (μ) to its spin (I) is called its **gyromagnetic** or **magneto-gyric ratio** (γ):

$$\gamma = \frac{\mu}{I} \quad or \quad \mu = \gamma \cdot I \qquad (2\text{-}1)$$

"Gyro" refers to the proton's spin, while "magneto" refers to its magnetic moment.

In the next few chapters we will see that nuclear spin and nuclear magnetic moment are both essential components of nuclear magnetic resonance. The gyromagnetic ratio just discussed will play a central role in determining the frequency of the radio waves that are required to make magnetic resonance occur.

3. Simple Quantum Model of Nuclear Spin: Parallel & Antiparallel Protons

In magnetic resonance we observe a signal produced by the magnetic moments of the protons. This signal is an electric current induced in a receiver coil by the magnetic moments. The magnetic moment of a single proton is, however, far too small to induce a detectable current in a coil, so in some way, the protons must be made to work together to produce a larger, detectable net magnetic moment in the body.

Normally, protons in the body have a completely random orientation. Their tiny magnetic moment vectors point in all directions and cancel out so that no net magnetic moment is produced. However, when the protons are placed in a strong, static magnetic field, the orientations of their magnetic moments or spins are no longer totally random. They will preferentially point in the direction of the magnetic field. This cooperation of a large number of protons creates a net magnetic moment that can produce detectable signals.

The effect of the static magnetic field of an MR magnet on the orientations of the proton spins can be explained by two rather different models. Just as different aspects of the physics of light can be explained either by the classical wave theory or by the quantum particle theory, the principles of MR can be described either by classical physics or by quantum mechanical theories.

On the surface, the wave and particle theories of light seem very contradictory. It takes a rather involved physics theory to show how they are in fact consistent, in a sense just two faces of the same coin. In the same way, the classical and quantum theories of MR seem very different at the level at which we will discuss them. Thus it is important not to mix the two models in any particular explanation.

However, in a more complete theory of MR the true relationship between the classical and quantum views of MR becomes apparent, and the differences between them can be reconciled. In this book we will discuss both the simple classical and quantum models since readers are likely to encounter both models in their reading; also, certain MR phenomena can be easier to visualize in one or the other model.

In the simplified quantum model the magnetic moments of the protons in a static magnetic field can have only two possible orientations: either **parallel** or **antiparallel** to the static magnetic field direction (Fig. 3-1).* The two orientations represent two different **energy levels** of the proton (Fig. 3-2). A parallel proton has a slightly lower energy content than an antiparallel proton. As is common to most phenomena in nature, the lower energy state is preferred. Therefore, when a collection of protons is placed in a static magnetic field, the parallel protons will, after a short while, outnumber the antiparallel protons. Due to the surplus number of parallel protons, a net magnetic moment vector is created. The net magnetic moment vector that results in a unit mass of tissue is

---

* Strictly speaking, in the complete quantum theory the magnetic moment of the proton does not in fact point directly along the magnetic field direction. Rather, it exists along either of two cones centered along the magnetic field direction (corresponding to the parallel and antiparallel orientations).

Oriented on such a cone, the magnetic moment always maintains a constant angle with the magnetic field direction. Its position on the cone, however, is *completely random*, a manifestation of the inherent uncertainty that is an essential characteristic of the complete quantum theory. Because of the random motion of the magnetic moment along the cone, the components of the magnetic moment in the plane perpendicular to the magnetic field direction average zero; therefore, the average value of the magnetic moment, or its **expectation value**, as physicists call it, points directly along the magnetic field direction in either a parallel or antiparallel orientation. It is this average or expectation value of the magnetic moment in quantum theory that is the useful concept for *all* of our discussions in this book. *(footnote continues on next page)*

referred to as the **magnetization vector** or simply as the **magnetization, M** (Fig. 3-1).

The creation of a magnetization vector after the protons are placed in a magnetic field is one manifestation of a process called **T1 relaxation, longitudinal relaxation,** or **thermal relaxation**. This process will be discussed in detail in Chapters 8 and 9.

When the surplus number of parallel protons has reached its maximum, and thus when **M** is a maximum and no longer changing, the parallel and antiparallel protons are said to be at **thermal equilibrium**. It is this magnetization vector that, by some manipulation, will induce a current in the receiver coil and produce the MR signal.

At thermal equilibrium, the ratio of the number of parallel ($N_{parallel}$) to the number of antiparallel ($N_{antiparallel}$) protons in a unit mass of tissue is given by the **Boltzmann equation.** (Physicists usually refer to this equation as the **Boltzmann distribution.**) In the equations that follow, ΔE is the difference in energy content between an antiparallel and a parallel proton, k is the Boltzmann constant, and T is the absolute temperature

---

(cont. from previous page) We recommend that the reader visualize the two allowed quantum states of the proton in a magnetic field as having magnetic moments that point directly along the magnetic field direction, either in the same direction (parallel) or in the opposite direction (antiparallel). Strictly speaking, the magnetic moment in this visualization is the expectation value of the magnetic moment. At the level of explanation used in this book and other introductory MR texts written for non-physicists, the introduction of the actual orientation of the magnetic moment along two fixed cones, rather than simply using the expectation value of the magnetic moment, is generally confusing and non-productive. We mention it here simply because it is, in fact, used in many such introductory texts.

In the simple quantum theory, the proton can exist in either of two states. These two states correspond to two allowed orientations for the expectation value of the magnetic moment: either directly parallel or antiparallel to the magnetic field. In the classical theory we will see that the magnetic moment can point in any direction; it is not restricted to only two orientations.

In the complete quantum theory it is possible to form combinations or "superpositions" of the two allowed states that allow the expectation value of the magnetic moment to point in any direction. This complete quantum theory is beyond the scope of our discussions in this book. We simply note here that many of the apparent contradictions between the simple quantum and classical theories can be resolved by using the complete theories with their full complexity. In fact, the classical description of the motions of the magnetic moment and the complete quantum theory description of the motions of the expectation value of the magnetic moment yield the same equations!

measured in degrees kelvin (°K). kT is approximately the average thermal energy contained by each of the individual molecules due to its random thermal motions.

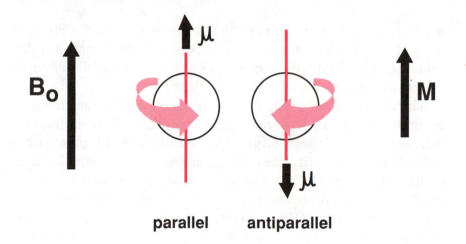

Fig. 3-1. The simple quantum model of proton spin. When protons are placed in a static magnetic field (B$_0$) their magnetic moments (μ) are allowed to point in only two directions: either parallel or antiparallel to the magnetic field. At thermal equilibrium the number of protons in the parallel orientation will be greater than the number in the antiparallel orientation: as a result, the net magnetic moment vector per unit mass — the magnetization vector (M) — will point in the same direction as B$_0$.

Energy Levels

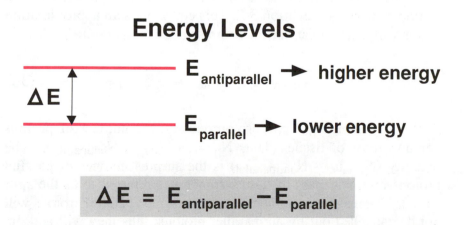

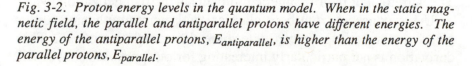

Fig. 3-2. Proton energy levels in the quantum model. When in the static magnetic field, the parallel and antiparallel protons have different energies. The energy of the antiparallel protons, E$_{antiparallel}$, is higher than the energy of the parallel protons, E$_{parallel}$.

The Boltzmann equation:

$$\frac{N_{parallel}}{N_{antiparallel}} = e^{\Delta E/kT} \qquad (3\text{-}1)$$

where e is the base of the natural logarithms (approximately 2.7). (If you are uncomfortable with exponential functions a fuller explanation of them will be given in Chapter 8. A simpler approximation to Equation 3-1 will be given shortly.)

The Boltzmann equation (3-1) tells us that as the absolute temperature approaches zero, the ratio of parallel to antiparallel protons becomes infinitely large. This means that all protons are parallel; in other words, they are all aligned in the same direction as the static magnetic field and are all in the lower energy state. At temperatures above absolute zero, some protons will be in the higher energy state. These protons are "bumped" into the higher energy state by the thermal motions of the molecules of the tissue. The higher the temperature, the greater the thermal motions, and the larger the fraction of protons in the higher energy state. With temperatures at or above room temperature, ΔE is much, much less than kT. Thus $\Delta E/kT$ is nearly zero and $e^{\Delta E/kT}$ is nearly equal to one ($e^0 = 1$). This means that at these temperatures, the number of protons in the higher and lower energy states will be almost equal.

Using the fact that ΔE is very much less than kT at temperatures at or above room temperature, we can transform Equation 3-1 into a simpler form — Equation 3-2. This equation is an approximation that is very accurate at normal room or body temperatures.

$$N_{parallel} - N_{antiparallel} = \frac{\Delta E}{2kT} N_p \qquad (3\text{-}2)$$

In this equation, N_p is the proton density: the number of protons per unit mass of tissue. (Thus $N_p = N_{parallel} + N_{antiparallel}$.) The quantity ($N_{parallel} - N_{antiparallel}$) is the surplus number of parallel protons in a unit mass of tissue, sometimes referred to as the **spin excess**. The magnetic moments of these surplus parallel protons will not be cancelled out by antiparallel protons; thus they will add up. The sum total of their magnetic moments is the magnetization, **M**.

Using concepts from both classical and quantum physics, and a bit of mathematics, we can determine the actual size of ΔE. This derivation is not particularly interesting for our purposes so we will only quote the result, which is of interest. ΔE is found to be directly proportional to the strength of the static magnetic field:

$$\Delta E = \hbar\,\gamma\,B_0 = \frac{h}{2\pi}\,\gamma\,B_0 \qquad (3\text{-}3)$$

where $\hbar$ is a constant of nature called Planck's Constant*, and γ is the gyromagnetic ratio of the proton. B_0 is the magnetic field strength. The units used for magnetic field strength are **tesla** (T) and **gauss** (G). 1 gauss is equal to 10^{-4} tesla. The magnetic field strength of Earth is about 3×10^{-5} T (0.3 G) at the equator and about 7×10^{-5} T (0.7 G) at the north pole. The majority of the magnets used for MRI have strengths thousands of times greater than the earth's field, in the range of 0.1-1.5 T (1,000-15,000 G).

Using the formula for ΔE from Equation 3-3, we can rewrite Equation 3-2 as:

$$N_{parallel} - N_{antiparallel} = \left(\frac{\hbar\,\gamma}{2\,k}\right)\frac{B_0\,N_p}{T} \qquad (3\text{-}4)$$

Like Equation 3-2, the above equation is an excellent approximation as long as the temperature is not too close to absolute zero — e.g., near room or body temperatures the equation works just fine. The quantity in brackets is a constant for protons and is equal to $0.001°$ K/T.

Equation 3-4 says that, *at thermal equilibrium, the surplus number of parallel protons in a unit mass of tissue, and therefore the size of the magnetization vector, **M**, is directly proportional to the strength of the static magnetic field and to the proton density, and inversely proportional to the absolute temperature.* This means that the magnetization increases with increasing magnetic field strength or proton density, and decreases with increasing temperature:

$$\text{at thermal equilibrium,} \quad M = c \cdot \frac{B_0 N_p}{T} \qquad (3\text{-}5)$$

---

* The term "Planck's Constant" is often used to refer to either of two slightly different numbers which are represented by the following two notations: $\hbar$ and h. Just to keep things straight, $\hbar = h/2\pi$. Similarly, $\gamma/2\pi$ is sometimes written as γ.

In Equation 3-5, "c" is a constant for protons.

Using Equation 3-4, we find that *at body temperature and with the range of field strengths used in MRI, the surplus number of parallel protons is only 0.3 to 5 protons per 1 million protons.* It may seem strange that such a small surplus number of parallel protons can create a magnetization vector sufficiently large to induce a detectable current in a receiver coil. However, the magnetization vector created in a volume of tissue will also be proportional to the total number of protons present in that volume, i.e., to the proton density in the tissue. In only 1 ml of water, there are about 6×10^{22} protons. It is the great abundance of protons in human tissue (both in water and fat) that makes MRI feasible.

4. Classical Model of Nuclear Spin: Precessional Motion, Larmor Frequency

As mentioned previously, the behavior of protons in a static magnetic field can be explained by two different models. The simple quantum model described in the previous chapter explained how the distribution of parallel and antiparallel protons leads to the formation of a magnetization vector. The mathematics in that model says that the size of the magnetization vector is proportional to both the strength of the static magnetic field and to the proton density. It will now be useful to turn to the classical model to see how this magnetization vector can change its orientation.

According to classical physics, if a magnet is placed into a magnetic field, the magnetic field will try to turn the magnet so that its magnetization vector aligns with the direction of the magnetic field. We have all observed this phenomenon when we have used a compass: when it is free to move, the compass needle, which is a magnet, aligns with the earth's magnetic field. This twisting force produced by the magnetic field of the earth is called a **torque**.

A magnetic field will exert a torque on a magnet to try to twist its magnetization into parallel alignment with the magnetic field.

Torques can be produced by forces other than magnetic forces. For example, if a top that is not spinning is placed upright on a table, it will fall over. In this case, gravity produces a torque which rotates the top from a vertical to a horizontal position. If, however, the top is spinning, a very interesting phenomenon occurs. The torque due to gravity no longer simply pulls the top over. Instead, the top **precesses**: its spin axis moves in a cone around the vertical direction — the direction of gravity.

*If a torque is exerted on a spinning object, its spin axis will not move in the same direction as the twisting produced by the torque. Instead, the spin axis will move in a direction perpendicular to the torque's twisting. The result is a motion called **precession**.*

A proton has *both* a magnetic moment *and* a spin. Suppose a proton is placed into a static magnetic field such that its spin axis (or its magnetic moment vector) makes an angle with the direction of the magnetic field. The magnetic field will then exert a torque on the proton. This torque will try to align the proton with the direction of the magnetic field. Since the proton is also spinning, however, the torque will not twist the magnetic moment vector (μ) into alignment with the magnetic field. Rather, μ will precess around the magnetic field direction while maintaining a constant angle with it (see Fig. 4-1).

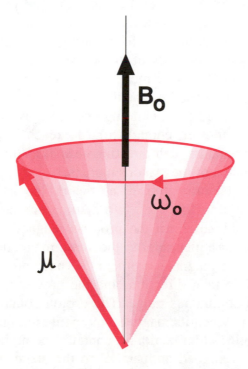

Fig. 4-1. Precession of the proton. The torque exerted on the magnetic moment of the proton (μ) by the static magnetic field (B_0) causes μ to precess around B_0 with an angular frequency of ω_0.

In reality, the angle of the cone of precession of the proton will change with time (very slowly compared to the speed of the precessional motion). One cause is analogous to the motion of a spinning top. As the top precesses, it slowly falls over due to effects of friction with the table. This friction slowly takes energy away from the spinning top. In a similar way, interactions of the proton with the other protons and atoms around it can take energy away from the proton. This is the same process of T1 relaxation which was briefly mentioned in Chapter 3.

If the temperature of the tissue containing the proton approached absolute zero, the proton would eventually lose all of its energy and align exactly with the magnetic field. At higher temperatures, the proton can also gain some energy from the thermal motions of the atoms, and will not exist in a state of constant alignment with the magnetic field.

At this point another analogy will provide a useful mental picture for the protons in the tissue. Imagine a group of compass needles that are continuously jostled around. These compass needles will try to align with the magnetic field, but the jostling disturbs their orientation. At any instant the needles point in all directions, though there is some preference for pointing in the direction of the magnetic field. If the intensity of the jostling is reduced, this preference for alignment becomes more pronounced. In much the same way, the protons are jostled by the thermal motions of the atoms in the tissue and therefore do not all align perfectly with the magnetic field, even when allowed enough time to reach thermal equilibrium. As the temperature, and thus the thermal jostling, is reduced, the overall alignment of the protons increases.

In the simple quantum model, as the proton gains and loses energy it simply jumps between two distinct energy levels corresponding to the alignment of the proton's magnetic moment in a direction parallel or antiparallel to the static magnetic field. In the classical model, the magnetic moment can have any orientation to the static magnetic field direction, and the proton can contain any energy within a continuous range. The proton contains the lowest amount of energy when its magnetic moment is aligned parallel to the static magnetic field direction; it contains the highest amount of energy when it is aligned antiparallel to the static magnetic field direction.

The angular frequency of the precessional motion of the proton is directly proportional to the strength of the static magnetic field, B_0:

$$\omega_0 = \gamma B_0 \tag{4-1}$$

ω_0 is the **angular frequency** of the precession, and γ is the same gyromagnetic ratio encountered in Chapter 2. The reason for the presence of the gyromagnetic ratio in Equation 4-1 is easy to understand. The gyromagnetic ratio expresses the relationship between the magnetic moment and the spin of the proton. We have seen that the precession of the proton is a result of both its magnetic moment and its spin.

The angular frequency ω_0 in Equation 4-1 is measured in radians per second. There are 2π radians in a complete rotation. We can rewrite Equation 4-1 in terms of the precessional frequency (f_0) expressed in cycles per second or hertz (Hz). The relationship between ω_0 and f_0 is:

$$\omega_0 = 2\pi f_0 \tag{4-2}$$

Equation 4-1 then becomes:

$$2\pi f_0 = \gamma B_0 \quad or \quad f_0 = \frac{\gamma}{2\pi} B_0 \tag{4-3}$$

The frequency of precession (f_0) is also called the **resonance frequency** or the **Larmor frequency**. Equation 4-1 or 4-3 is thus often referred to as the **Larmor equation**.

$\gamma/2\pi$ has units of frequency per magnetic field strength. For protons, $\gamma/2\pi$ is equal to 42.58 Mhz/T. This means that in a static magnetic field with a strength of 1 tesla, the protons will precess with a frequency of 42.58 Mhz. Since the gyromagnetic ratio of nuclei other than hydrogen is different, these other nuclei will precess at different frequencies.

You should be aware that in some MR literature, the terms γ and $\gamma/2\pi$ are confused. Often the term γ is mistakenly given the value 42.58 Mhz/T for the proton when the proper term for this value is $\gamma/2\pi$.

With all of the jostling due to thermal energy, the changes in orientation of the individual magnetic moments are extremely complicated. Luckily, we do not need to follow the motions of the individual protons to understand the MR phenomenon; rather, we can consider the net magnetic moment of all the protons. The motion of this magnetization vector, **M**, is relatively simple. If the magnetization vector is initially oriented at an angle to the static magnetic field direction, it will precess around the magnetic field direction with the Larmor frequency. Due to T1 relaxation, the magnetization, **M**, will also change the angle of its precession cone until it finally points exactly along the direction of the magnetic field when thermal equilibrium is reached.

5. Simple Quantum Model of Magnetic Resonance: Radio Waves & Photons

The magnetization vector, **M**, is a very important parameter in MRI, because, as stated in Chapter 3, it is the magnetization vector that, by some manipulation, will induce a current in the receiver coil and thus produce the MR signal. For the magnetization vector to induce a detectable current in the receiver coil, its direction must vary in time so that the vector, or at least a component of it, is alternately pointing into and out of the coil. At thermal equilibrium, the magnetization vector created by the protons in a static magnetic field is constant in size and direction (parallel to the magnetic field) and therefore cannot induce a current in the receiver coil. To be capable of inducing a signal current, the magnetization must be moved away from its static parallel position. This means that the normal equilibrium between the two proton populations must be disturbed.

The simple quantum model is not useful for explaining the motion of the magnetization vector, but it does explain how the equilibrium between the parallel and antiparallel protons can be disturbed. As explained in Chapter 3, when a proton is placed in a static magnetic field, two separate energy levels are created, corresponding to the parallel and antiparallel orientations of the proton magnetic moment. The proton can exist in either level or orientation. Because the lower energy state is slightly preferred, there will always be a surplus of protons with their spins oriented in the parallel direction at thermal equilibrium.

This equilibrium is not static. Although the overall ratio of parallel to antiparallel protons remains constant, the individual protons jump back and forth between the two energy states by exchanging energy with each other and with the energies of the molecular thermal motions (Fig. 5-1). At thermal equilibrium the number of transitions from the higher to lower energy state is equal to those from the lower to higher energy state. Thermal equilibrium can be disturbed by adding energy to the collection of protons in the tissue, producing more transitions from the lower to higher energy state than from the higher to lower state. This can be accomplished by irradiating the protons with radio waves of a precise frequency.

To understand this central process in magnetic resonance we need to delve a bit into the nature of radio waves. Radio waves are a form of **electromagnetic (E-M) radiation**. Other forms of E-M radiation include microwaves, infrared, visible light, ultraviolet, x-rays, and gamma-rays. All E-M radiation is essentially an oscillating electric and magnetic field traveling through space, otherwise known as a traveling wave. It is analogous to a water wave traveling over a pond. A water wave is a propagating disturbance in the water surface. If you observe the water at a fixed position in the pond, the water surface will move up and down (or oscillate) as the wave passes by. If you observe the electric or magnetic field at a fixed point in space as E-M radiation passes that point, you will detect an oscillation of the field. The differences between the various types of E-M radiation lie in the frequencies at which these fields oscillate. Radio waves have a relatively low frequency while x-rays and gamma-rays have very high frequencies.

Energy Levels

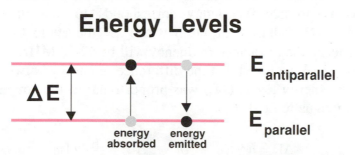

5-1. In the simple quantum model, the transformation of the proton's magnetic moment from a parallel to an antiparallel orientation involves an absorption of energy equal to ΔE, while the transformation from an antiparallel to a parallel orientation involves the emission of the same amount of energy ΔE. One means by which the proton can absorb or emit energy is by the absorption or emission of photons of the correct energy (ΔE).

This description of E-M radiation as a traveling wave is essentially a classical physics explanation. However, E-M radiation has other properties that can best be visualized by considering it as composed of particles rather than as a wave. In the quantum physics model, E-M radiation is composed of discrete packets of energy called **photons**. The quantum model describes the emission and absorption of E-M radiation energy as the creation and absorption of photons. The energy (E) contained in these photons is determined by the frequency (f) of the radiation:

$$E = hf \qquad (5\text{-}1)$$

where h is Planck's constant.

If we irradiate protons in a static magnetic field with E-M radiation, we might induce some of the protons to absorb energy from this radiation and to jump from the lower to the higher energy state. This will only occur, however, if a very special condition, the **resonance condition**, is satisfied. This condition states that the energy of the photons comprising the radiation (hf) must exactly equal the difference in energy between the two energy levels of the proton (ΔE):

$$\Delta E = hf \qquad (5\text{-}2)$$

For protons in the static magnetic field produced by an MRI scanner, this **resonance frequency** corresponds to that of radio-frequency E-M radiation or radio waves. If the static magnetic field is 1 tesla, the resonance frequency will be 42.58 MHz.

In Chapter 3 we noted that the difference in energy between the two proton energy levels (ΔE) was proportional to the strength of the static magnetic field (B_0):

$$\Delta E = \hbar \gamma B_0 \qquad or \qquad \Delta E = \frac{h}{2\pi} \gamma B_0 \qquad (5\text{-}3)$$

If we substitute this value for (ΔE) into Equation 5-2, we obtain:

$$\frac{h}{2\pi} \gamma B_0 = hf \qquad or \qquad \frac{\gamma}{2\pi} B_0 = f \qquad (5\text{-}4)$$

Note that this frequency is identical to the Larmor frequency (f_0) in Equation 4-3. This means that *the radiation frequency required in the simple quantum theory to induce transitions of the proton from the lower to higher energy state is identical to the frequency of precession of the proton derived using the classical theory in Chapter 4*. In the next chapter we will see how the classical theory predicts that the precessional frequency is the frequency at which radio waves can interact with the protons and produce magnetic resonance. For this important result the classical and quantum theories agree exactly.

If a proton jumps from the higher energy state to the lower energy state (i.e., it changes from an antiparallel to a parallel orientation) it must give up an energy equal to ΔE, the difference between the two energy levels. One way it can give up this energy is by emitting a photon of frequency, $f = \Delta E/h$ as described in Equation 5-2. Whether the proton absorbs or emits a photon, the frequency of the photon involved is the same: $f = \Delta E/h$.

You might think that the irradiation of protons with radio waves (i.e., photons) of just the right frequency would lead to an ever increasing number of antiparallel protons and an ever decreasing number of parallel protons. That, however, is not the case. Two things may happen when protons are irradiated with photons having an energy equal to ΔE. If a parallel (lower energy) proton interacts with such a photon, then the proton can absorb the photon energy and flip to the antiparallel (higher energy) position. However, if an antiparallel (higher energy) proton interacts with the photon, then the proton can be induced to release energy in the form of another identical photon with energy ΔE, thereby flipping to the parallel (lower energy) position.

We can write the following two equations of energy exchange:

(a) 1 parallel proton + 1 photon = 1 antiparallel proton
(b) 1 antiparallel proton + 1 photon = 1 parallel proton + 2 photons

From the point of view of the proton, in situation (a), there is a net absorption of energy from a radio wave photon. In situation (b), there is a net release of energy into the creation of a radio wave photon. The net result in a collection of protons will depend upon the relative number of parallel and antiparallel protons when irradiated by the photons. This is shown in a very simplified manner in Figure 5-2. In Fig. 5-2a there is thermal equilibrium between the

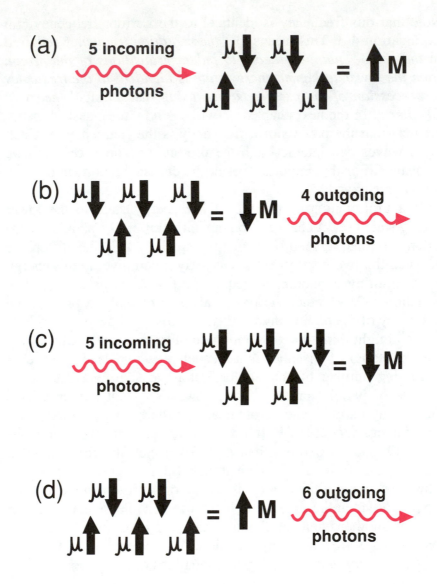

Fig. 5-2. Absorption and emission of photons by the protons and the resulting changes in the magnetic moments of the protons (μ) and in the magnetization (M). This is a very simplified example in which 5 protons each interact with one of the 5 incoming photons. Here, the protons are in a static magnetic field that in this figure points upward. In (a) and (b), 5 photons come in and 4 leave. All 5 protons change their state, but the net result is the conversion of one proton from the lower energy state to the higher energy state. In (c) and (d), 5 photons come in and 6 leave. All 5 protons change their state, but the net result is the conversion of one proton from the higher energy state to the lower energy state. The difference in the results is due to the difference in the initial states (a) and (c): in (a) more protons are in the lower energy state, while in (c) more protons are in the higher energy state.

two proton populations, with a surplus number of parallel protons and a magnetization vector (M) pointing in the parallel direction. The protons are then irradiated by photons having a frequency equal to the Larmor frequency of the protons. *The protons in the tissue have all the same likelihood of interacting with a photon, whether they have a parallel or antiparallel orientation.* Thus, let us assume that all the five protons shown in the figure do interact with a photon. Consequently, they all flip 180°. The result is a net absorption of the energy of a single photon, a surplus number of antiparallel protons and a net magnetic moment of equal size as before, but pointing in the antiparallel direction (Fig. 5-2b). However, if we now continue to irradiate the protons with photons (Fig. 5-2c), the surplus number of antiparallel protons will lead to a net release of energy and a return of the magnetization vector to the parallel orientation (Fig. 5-2d). For as long as the protons are irradiated, a net absorption followed by a net release of energy will occur, and the magnetization vector will flip back and forth between the parallel and antiparallel directions.*

This exchange of energy between protons and photons is the physical phenomenon named **nuclear magnetic resonance**.

Resonance can be defined as synchronous vibration; physically it entails the exchange of energy between two systems having the same natural frequency. We are all familiar with the resonance phenomenon that occurs whenever we use a radio. When we tune a radio we are changing the natural frequency of its electronics so that it matches the frequency of the radio waves broadcast by our favorite radio station. This tuning allows a transfer of energy from these radio waves into the radio so that we can listen to the selected station.

Exchange of energy between protons and photons in nuclear

---

* The astute reader may here realize that there is an apparent flaw in our reasoning. It is clear that if the magnetization originally points in the parallel direction, the absorption of energy from photons should be capable of equalizing the number of parallel and antiparallel protons and thus reducing the magnetization to zero. Once the number of parallel and antiparallel protons are equal in number, however, there appears no reason why their numbers should change further.

This apparent impasse is a limitation of our simple quantum theory. In the complete quantum theory, the magnetization is shown to simply rotate from a parallel to antiparallel orientation and back again without changing its size — it is never zero in this process. In fact, this motion of the magnetization is shown to be identical to the motion predicted using the classical model, as we will discover in the next chapter.

magnetic resonance can take place only when the frequency of the photons is equal to the Larmor frequency of the protons. That the resonance is "nuclear magnetic" means that the exchange of energy is dependent upon an interaction of the magnetic field of the radio wave with the magnetic moment of the hydrogen nucleus (Ch. 6). When referring to nuclear magnetic resonance imaging, the term "nuclear" is often omitted to avoid confusion with nuclear medicine and to avoid conjuring up the public's fears of nuclear radiation.

In summary: the simple quantum model explains the interaction between protons and photons as an exchange of energy. It also shows how the magnetic moments of the protons can be turned from a parallel to an antiparallel position, and vice versa, through this exchange of energy. It does not explain, however, how these magnetic moments can induce a current in a coil and produce the MR signal. For that we need to use the classical physics model.

6. Classical Model of Magnetic Resonance: 90° & 180° Pulses

In Chapter 4 we described the behavior of a proton in a static magnetic field using the classical model. Ignoring the effects of interactions with neighboring magnetic moments, if the proton's magnetic moment is initially oriented at an angle to the magnetic field direction, the magnetic moment will precess around the magnetic field direction. Due to the effects of the interactions with other protons and molecules, however, the motions of the magnetic moments of the individual protons are quite complicated as they are jostled about by the thermal energy. Luckily, the motions of the magnetization vector, **M**, are much simpler: since the magnetization is the sum of the proton magnetic moments in a unit mass of tissue, the random thermal fluctuations of the individual proton magnetic moments will be mostly averaged out. If the magnetization, **M**, is initially oriented at an angle to the magnetic field direction, the magnetization will precess around the magnetic field direction with the Larmor frequency:

$$\omega_0 = \gamma \, B_0 \qquad or \qquad f_0 = \frac{\gamma}{2\pi} \, B_0 \qquad\qquad (6\text{-}1)$$

The angle of the cone of precession will change in time as the magnetic moments of the protons experience a net, overall gain or loss of energy.

We can manipulate the magnetization by changing the angle of its cone of precession or, to put it another way, by changing the angle the magnetization vector makes with the magnetic field direction. This is accomplished by exposing the protons to radio waves of a particular frequency: the Larmor (or resonance) frequency. This is the precessional frequency of the proton (or of the magnetization vector). In Chapter 5 we saw that the simple quantum theory predicted that photons of this same frequency were required to transfer energy to or from the proton. In this essential way the classical and quantum theories agree exactly.

In the rest of this chapter we will use the classical theory. We will also, for the time being, ignore the effects of thermal interactions with the protons (T1 relaxation processes). This leaves radio wave interaction with the protons as the only way of changing the angle the magnetization makes with the magnetic field.

At this point it is necessary to define a coordinate system so that we can describe the interaction between the radio waves and the proton. Let's choose this coordinate system so that the static magnetic field, $\mathbf{B_0}$, points in the direction of the positive z-axis. Figure 6-1 shows this coordinate system along with the magnetization ($\mathbf{M}$) precessing around the static magnetic field ($\mathbf{B_0}$).

Remember from Chapter 5 that radio waves are electromagnetic radiation composed of oscillating electric and magnetic fields. It is the oscillating magnetic field* that will affect the orientation of the magnetization. To accomplish this desired effect, radio waves are transmitted into the patient in such a way that the oscillation of the magnetic field occurs in the x-y plane, i.e., in a plane perpendicular to the static magnetic field direction.

Like the static magnetic field, the oscillating magnetic field of the radio wave is also a vector: it has a strength and a direction. Let's assume that the oscillating magnetic field points alternately in the positive and negative directions along the y-axis. We can make a graph of the strength of the oscillating magnetic field as a function of time, at the position of the tissue element under study. This is done in Figure 6-2. Five different points in time are marked on this graph: *a*, *b*, *c*, *d*, and *e*. At times *a*, *c*, and *e*, the strength of the

---

* We will use the terms "oscillating magnetic field," "radio-frequency magnetic field," and "rf magnetic field" interchangeably.

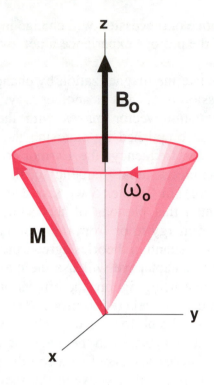

Fig. 6-1. The motion of the magnetization (M) in a static magnetic field (B_0): M precesses around B_0 with an angular frequency of ω_0 (expressed in radian/sec). $\omega_0 = 2\pi f_0$, where f_0 is the Larmor frequency expressed in cycles/sec or Hz. The x-y-z coordinate system described in the text is shown: the z-axis is defined as the direction of B_0.

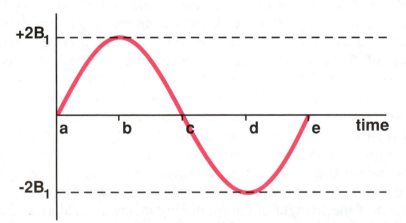

Fig. 6-2. The magnetic field of the radio wave that interacts with the proton This is a graph of the oscillating magnetic field of the radio wave as a function of time, at the position of the tissue element under study. The period of time from point a to point e is 1/(radio wave frequency). If the frequency is 10MHz, then the time period shown in this graph is 1/10,000,000 sec. The maximum strength of the magnetic field of the radio wave is $2B_1$.

oscillating magnetic field is zero. At time b, it is a maximum, and points in the direction of the positive y-axis. At time d, it is also a maximum but now points in the direction of the negative y-axis. We'll call the maximum value of the oscillating magnetic field $2B_1$. (We will see shortly why we wish to call it $2B_1$ rather than simply B_1.) Figure 6-3 shows the actual rf magnetic field vector at the same five points in time, a, b, c, d, and e, as in Figure 6-2.

Now comes a very important step. The oscillating magnetic field vector can be represented as the sum of *two* vectors. Though the original vector only points along the y-axis and varies in size, it can be considered to be made up of two vectors, each with a *constant* size of B_1, and both rotating in the x-y plane: one clockwise and one counterclockwise. This is demonstrated in Figure 6-4. Here the oscillating vector and its two rotating components are shown at the same times, a, b, c, d, and e, as in Figures 6-2 and 6-3. Let's label the clockwise rotating component "vector 1," and the counterclockwise rotating component "vector 2." At time a and c, vectors 1 and 2 point in opposite directions, completely cancelling; their sum is equal to zero. At time b, vectors 1 and 2 both point in the direction of the positive y-axis; their sum is $2B_1$ in the direction of the positive y-axis. At time d, vectors 1 and 2 both point in the direction of the negative y-axis; their sum now is $2B_1$ in the direction of the negative y-axis. If we were to continue to look at the sum of vectors 1 and 2 at all other possible times we would see that the sum of vectors 1 and 2 is always exactly equal to the oscillating magnetic field vector both in size and direction.

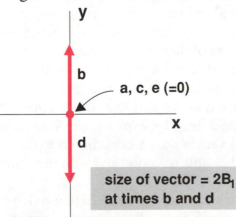

Fig. 6-3. The same radio wave magnetic field as described in Fig. 6-2. Here the magnetic field vector is shown in the x-y plane. At time b the vector is pointing in the positive y direction; at time d the vector is pointing in the negative y direction; at times a, c, and e the vector is zero.

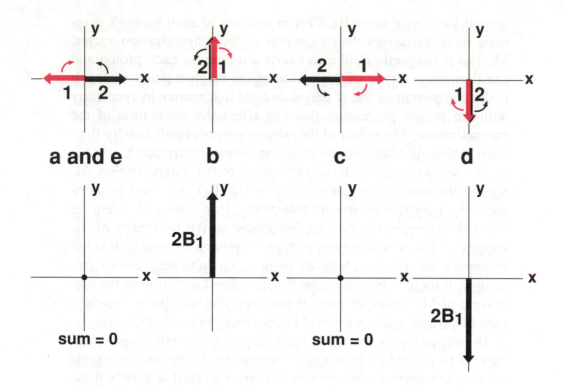

Fig. 6-4. The oscillating magnetic field of the radio wave represented as the sum of 2 vectors labeled 1 and 2. The same points in time — a, b, c, d, and e — are shown as in Fig. 6-2 and 6-3. The top figures show the vector components 1 and 2 while the lower figures show their sum, which is the actual oscillating magnetic field of the radio wave. Vectors 1 and 2 both have a constant size of B_1. Vector 1 rotates clockwise and vector 2 rotates counterclockwise.

The rotational frequency of the magnetic field vectors 1 and 2 is the same as the frequency of the radio wave — the Larmor frequency. Only one of these rotating magnetic field vectors has a significant effect on the magnetization — the one that is rotating around the z-axis in the same direction as the precession of the magnetization and in synchronization with this precession.

To conclude a lengthy explanation: The radio wave creates a magnetic field vector with a constant size (B_1), that rotates in the x-y plane, in the same direction and with the same frequency as the precession of the magnetization.

Looking back at Chapter 4 and Figure 4-1, you may recall that the static magnetic field (B_0) exerts a torque on the magnetic moment of the proton (μ) which tries to rotate μ into parallel alignment with B_0. Due to the spin of the proton, this torque produces a precession of the proton's spin axis and magnetic moment vector around the direction of B_0. This precession of each proton's magnetic moment causes the precession of the magnetization vector, **M**. The rf magnetic field also exerts a torque on each proton and can thus cause a type of precession of the magnetization vector. It is the component of the rf magnetic field that rotates in synchrony with the proton precession that can affect the orientation of the magnetization. The effect of the other component will rapidly fluctuate in time and have no net effect on the magnetization.*

In order to visualize this "precession" of the magnetization, **M**, around the rotating magnetic field vector (B_1), we need to introduce the **rotating frame of reference**. This frame of reference looks the same as the one we introduced at the beginning of the chapter with x, y, and z axes, with one essential difference: it is not stationary but rotates about its own z-axis with the Larmor frequency. It rotates with the same frequency and direction as the precession of **M**; looked at from above (from the positive z direction) both this frame and the vector **M** move in a clockwise direction.

In the stationary frame of reference, the magnetization vector appears to be rapidly moving, or precessing. In the rotating frame the magnetization vector appears stationary since it is simply moving along with the rest of the frame. This effect can be compared to that experienced on a carousel. In the stationary frame of reference, you stand on the ground next to the carousel watching the horses (or swans, or whatever) both rotate with the carousel and move up and down. In the rotating frame of reference, you stand on the carousel and rotate along with it. The rotational motion of the horses is then eliminated, relative to you, and all you then observe is their up and down motion.

In the rotating frame, since the x and y axes are moving, let's label them x' and y' to distinguish them from the axes in the stationary frame. Suppose that the magnetization, **M**, is oriented at some arbitrary angle to the z-axis while a radio wave at the Larmor frequency is transmitted into the patient in such a way that B_1 points along the positive y direction (Figure 6-5). In the rotating frame, B_1 will not move. However, **M** will precess around B_1 in the rotating frame in the same manner that it precesses around B_0 in the

---

* In the rest of this book we will use the following convention: We will refer to the static magnetic field simply as B_0, while the component of the rf magnetic field that rotates in the same direction as the precession of the magnetization will be referred to as simply B_1.

stationary frame. The precession of $\mathbf{M}$ around $\mathbf{B_1}$ will be governed by the same equation that governed its motion around $\mathbf{B_0}$:

$$\omega_1 = \gamma\, B_1 \qquad\qquad (6\text{-}2)$$

This equation is identical to Equation 6-1 except that now we are describing ω_1, the precessional frequency of $\mathbf{M}$ around $\mathbf{B_1}$ (expressed in radians per second). Since $\mathbf{B_1}$ is generally many thousands of times smaller than $\mathbf{B_0}$, the precessional frequency of $\mathbf{M}$ around $\mathbf{B_1}$ will be *much* slower than the precession around $\mathbf{B_0}$.

If we now look at the net motion of $\mathbf{M}$ in the stationary frame we see the results of precession around both $\mathbf{B_1}$ and $\mathbf{B_0}$. $\mathbf{M}$ appears to precess around $\mathbf{B_0}$ while the angle of the cone of precession changes in a cyclic fashion. The tip of $\mathbf{M}$ will trace out a spiral as it both precesses around $\mathbf{B_0}$ and changes its angle with the z-axis as a result of its precession around $\mathbf{B_1}$ (Figure 6-6).

Since the magnetization, $\mathbf{M}$, is a vector, we can talk of it as being composed of two components that can be added together to give the total vector. The component of magnetization along the z-axis will be called $\mathbf{M_z}$, while the component that lies in the transverse (x-y) plane will be called $\mathbf{M_{xy}}$. As shown in Fig. 6-7 any magnetization vector of any orientation can be viewed as the sum of its $\mathbf{M_z}$ and $\mathbf{M_{xy}}$ components. If the magnetization points directly along the z-axis, then it only contains the $\mathbf{M_z}$ component; while if the magnetization points in the x-y plane, it only contains the $\mathbf{M_{xy}}$ component. Often a magnetization that points in the z direction is simply labeled $\mathbf{M_z}$, while one that points in the x-y plane is labeled $\mathbf{M_{xy}}$.

Suppose that we first allow the protons in the tissue to reach thermal equilibrium while applying only the static magnetic field. The small excess of proton magnetic moments pointing parallel to $\mathbf{B_0}$ will produce a magnetization pointing in the same direction as $\mathbf{B_0}$ (in the positive z direction). Let us now transmit a radio wave at the Larmor frequency into the tissue so that $\mathbf{B_1}$ points along the direction of the positive x'-axis in the rotating frame (Fig. 6-8). Viewed in the rotating frame, $\mathbf{M}$ will then precess around the $\mathbf{B_1}$ direction so that it rotates from parallel to $\mathbf{B_0}$, down into the x'-y' plane, further down into the negative z direction (antiparallel to $\mathbf{B_0}$) and then back up into the x'-y' plane and further up into the positive z direction (parallel to $\mathbf{B_0}$) and so on for as long as the radio wave is transmitted. In the stationary frame, the tip of the magnetization

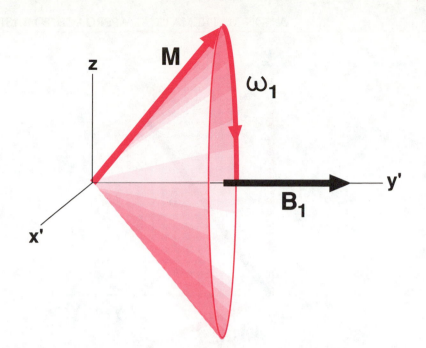

Fig. 6-5. The precession of the magnetization (M) around the rotating magnetic field vector (B₁) of the radio wave. This is shown in the rotating frame of reference. The frequency of the precession: $\omega_1 = \gamma B_1$. In this figure, M was not pointing in the z direction when the radio wave started. The darker section of the circle traced out by the tip of M corresponds to the spiral traced out by the tip of M in Fig. 6-6.

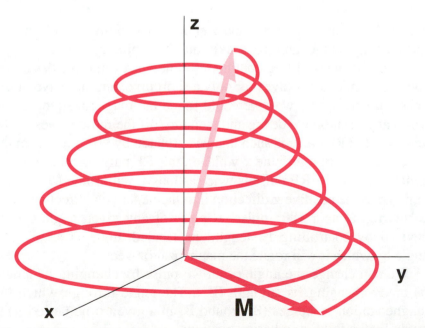

Fig. 6-6. The motion of the magnetization (M) shown in the stationary frame of reference during radio wave transmission. The part of the spiral motion of the tip of M in the stationary frame that is shown here corresponds to the darker section of the circle traced out the tip of M shown in Fig. 6-5. Figures 6-5 and 6-6 are two ways of representing the same physical process, one using a rotating frame of reference, the other using a stationary frame.

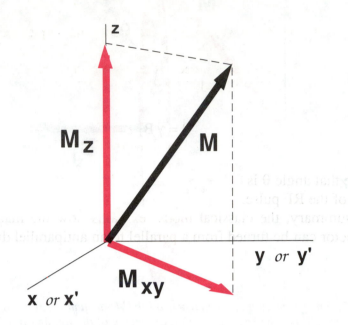

Fig. 6-7. A magnetization vector pointing in any direction can be considered to be composed of a component along the z-axis (labeled M_z) and a component in the x-y plane (labeled M_{xy}).

will start on the positive z-axis, spiral down into the x-y plane, further spiral to the negative z-axis and then spiral back up.

Instead of transmitting continuous radio waves into the tissue, we could transmit radio waves for only a short time, a radio wave pulse, or **RF pulse**. In this way we could rotate the magnetization vector, **M**, to any position we desire and then turn off the radio waves. If the length of the RF pulse is such that **M** is rotated by 90° (e.g., from the positive z direction into the x-y plane), this RF pulse is referred to as a **90° pulse**. If the RF pulse is twice as long it will rotate **M** by 180° (e.g., from the positive z direction into the negative z direction); this is referred to as a **180° pulse**. A pulse of any given angle can be devised by controlling its length; its labeled angle is simply the angle by which it will rotate the magnetization vector.

We can change the angle of an RF pulse by changing its length (t) or by changing its strength (B_1). The angle through which the magnetization precesses (θ) around B_1 in a given time period (t) is given by the following basic equation:

$$\theta = \omega_1 t \qquad (6\text{-}3)$$

where ω_1 is the frequency of precession around $\mathbf{B_1}$ expressed in radians per second. By substituting this into Equation 6-2 we obtain the relationship between the total angle of precession around $\mathbf{B_1}$ and the length and strength of the RF pulse:

$$\theta = \gamma\, B_1\, t \tag{6-4}$$

We see that angle θ is directly proportional to both the strength and length of the RF pulse.

In summary, the classical model explains how the magnetization vector can be turned from a parallel to an antiparallel direction

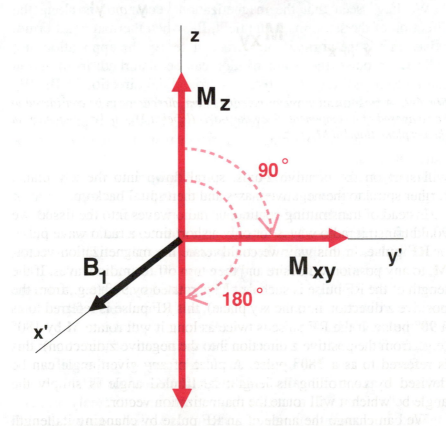

Fig. 6-8. The effects of a 90° pulse or a 180° pulse on the magnetization vector as shown in the rotating frame of reference. *The rotating magnetic field due to the radio wave (B_1) appears fixed in the x'-y' plane; in this figure it is shown along the x'-axis. M will then precess around B_1 as indicated; M will remain in the y'-z' plane. After the application of a 90° pulse, M will appear fixed along the y'-axis in the rotating frame. After application of a 180° pulse, M will appear fixed along the z-axis.*

by applying a 180° pulse. It also explains how the magnetization vector can be tilted into the x-y plane, perpendicular to the static field direction, by applying a 90° pulse; the magnetization will then rotate or precess in the x-y plane. This precession of the magnetization vector at the Larmor frequency forms the basis for the production of the MR signal.

7. The Magnetic Resonance Signal, Free Induction Decay (FID)

We have seen that the magnetization vector points along the direction of the static magnetic field, B_0, when thermal equilibrium exists; this is the normal undisturbed state. By the application of a radio wave pulse, the magnetization can be disturbed; its direction can be changed so that it precesses around the direction of B_0. We now want to look at how this precession of the magnetization vector produces a signal in the receiver coil. This signal provides us with the means of obtaining information from the magnetic resonance process.

Starting from a condition of thermal equilibrium, we will induce magnetic resonance in some tissue under study. The application of a 90° pulse twists the orientation of the magnetization vector, M, by 90°: before the pulse, M points in the z direction, and immediately after the pulse it points in the transverse (x-y) plane. The magnetization vector will then rotate or precess in this transverse plane. If a receiver coil is placed around or close to the tissue under study and is oriented so that its axis is aligned in the transverse plane (Fig. 7-1), the magnetization vector will alternately point into and out of the coil. This variation in time of the component of the magnetization along the coil axis induces an electric current in the coil which is detected as a signal called **free induction decay** or **FID**.

"Free" refers to the fact that the magnetization is freely precessing and no longer experiencing the torque produced by the RF pulse. "Induction" indicates that the current was produced using the principle that a changing magnetic field within a closed coil will induce an electric current in the coil (in this case the changing magnetic field is produced by the changing magnetization of the tissue). "Decay" means that this signal decreases with time for reasons we will shortly discover.

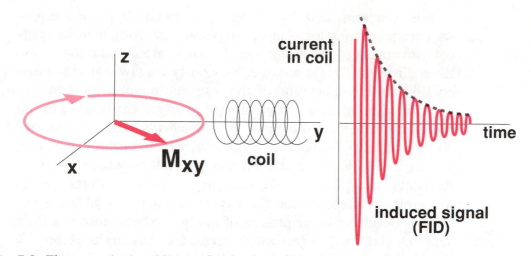

*Fig. 7-1. The magnetization (M) immediately after a 90° pulse will precess in the x-y plane of the stationary frame of reference. A receiver coil with its axis in the x-y plane will pick up a signal from this precessing magnetization. The electric current induced in the coil will oscillate at the Larmor frequency; its amplitude will decrease with time as the magnetic moments of the individual protons dephase. This oscillating signal in the coil is called the **free induction decay** or **FID**. Some authors use the term FID to refer instead to the amplitude or "envelope" of the oscillating signal; this is shown in the figure as a dashed line.*

The induced current in the receiver coil produces an oscillating signal that has the same frequency as the precessing magnetization (the Larmor frequency). The amplitude of this oscillating signal decreases with time, typically over a period of milliseconds. Usually the oscillating signal itself is called the FID. However, some authors use the term FID to refer to the amplitude or "envelope" of the oscillating signal (Fig. 7-1). This inconsistency in terminology is seldom confusing since it is usually obvious from context whether the author is referring to the oscillating signal or to its amplitude.

So far we have indicated no way in which the magnetization from different parts of the tissue can be differentiated, since they all precess at the same frequency and produce signals at the same frequency. If, after the application of the 90° pulse, we modify the static magnetic field, B_0, so that it varies slightly at different points in the body, then the magnetization in those parts of the body will precess at slightly different frequencies. The signal induced in the coil will then be composed of many slightly different frequencies, each one giving information about different locations within the tissue being studied. This is the basic means by which magnetic resonance imaging is accomplished.

It is commonly said that immediately after a 90° pulse, the protons are precessing together or "in phase," implying that the magnetic moments of the protons in the tissue are all pointing in the same direction. This is not true. Previously we saw that in thermal equilibrium the orientation of the magnetic moments was mostly random with only a slight preference to alignment parallel to the static magnetic field direction. If we mentally remove from our consideration all the protons whose magnetic moments cancel out, we are left with only a few protons per million whose magnetic moments point parallel to the static magnetic field. We referred to this group of protons as the spin excess. It is quite useful conceptually to consider only the protons of the spin excess, since it is their sum that yields the magnetization vector and thus produces the MR signal. After a 90° pulse we can think of the protons in this spin excess as precessing together in phase.

You might think that after a 90° pulse, these spin excess protons could just keep on precessing together in phase, so that M_{xy} remains constant. This would be true if all the protons experienced exactly the same static magnetic field. In fact, the protons do not all experience the same magnetic field; there are slight variations and fluctuations in the field due to the unavoidable limitations of the equipment producing the magnetic field, and due to the small, varying magnetic fields produced by the molecules in the tissue. In the imaging process, intentional variations of the magnetic field are also introduced.

Since each proton experiences its own slightly different magnetic field, each proton will precess at a slightly different frequency. As the individual magnetic moments of these protons dephase, i.e., spread out in several directions within the x-y plane, the size of the magnetization vector in that plane decreases, causing a decrease in the amplitude of the induced electric current (Fig. 7-1). When the protons comprising the spin excess are completely dephased, i.e., when they are evenly distributed in all directions within the x-y plane, there is no longer a magnetization vector in the x-y plane and thus no longer an MR signal. At this point we have reached the end of the FID signal.

The MR signal used for image reconstruction can be the FID itself. As will be discussed later, it is more common to use the so-called "echo" of the FID for imaging (Ch. 13), but for didactic purposes, we will initially assume that we use the FID for image reconstruction. The image itself is formed by a computer and is composed of many small picture elements or **pixels**, each corresponding to a particular volume element or **voxel**, in the slice of tissue being imaged. Each pixel is given a brightness or shade of gray determined by a mathematical processing of the amplitudes of the FIDs (or echoes) induced by the magnetization vector in the corresponding voxel. There may be 256 x 256 pixels in one image, so obviously a single FID (or echo) cannot contain enough information for the reconstruction of an image. Exactly how each pixel can be assigned its proper brightness will be discussed in Chapter 18; for the time being it suffices to know that several FIDs, and hence several 90° pulses, are needed before an image can be reconstructed.

The differences in the amplitude of the FIDs induced by the various tissues within the slice being imaged determines the contrast in the image. This contrast is highly dependent upon the time interval between each 90° pulse transmission. To understand why, we need to know what happens to the protons between the times at which the RF pulses are applied.

The first process to be considered is called longitudinal (or T1) relaxation.

II. Relaxation & Image Contrast

8. T1 Relaxation, The Exponential Function

If a sample of tissue is placed in a static magnetic field that points in the positive z direction, after thermal equilibrium is reached the magnetization due to the protons in the tissue will also point in the positive z direction. The size of this **equilibrium magnetization** is called M_0 and is given by Equation 3-5.

If a 90° pulse is then applied, the magnetization will rotate into the transverse (x-y) plane, and there will be no component of magnetization remaining along the z-axis. The protons will now begin the process of reestablishing the original condition of thermal equilibrium.

This thermal equilibrium has two fundamental properties: *(1) no component of magnetization in the transverse plane ($M_{xy} = 0$)*; and *(2) a surplus number of parallel oriented protons, yielding a magnetization along the z-axis equal to the equilibrium magnetization (M_0) prior to the RF pulse.*

Thermal equilibrium is regained by two different processes, corresponding to the above two properties (Fig. 8-1). The faster of these processes involves a loss of transverse magnetization due to dephasing of the protons; this promotes equilibrium property (1). We saw this effect in the last chapter; it causes the signal decay of the FID. This dephasing process has two distinct causes, one due to imperfections in the static magnetic field produced by the MR magnet and the other due to the inherent properties of the tissue (Ch. 12).

The reestablishment of a magnetization equal to M_0 along the z-axis — in the longitudinal direction — is a slower process which promotes equilibrium condition (2). This process is called **T1 (or longitudinal) relaxation**.

The growth of the magnetization along the z-axis after a 90° pulse is shown by the T1 relaxation curve in Fig. 8-2. The magnetization (M_z) increases from zero with an ever decreasing rate. **T1** is the time constant of the exponential function describing the curve. After a time equal to T1, 63% of the original magnetization (M_0) has been restored. After a time of 5 x T1, 99.3% of M_0 is reestablished. A short T1 thus means a fast restoration of the

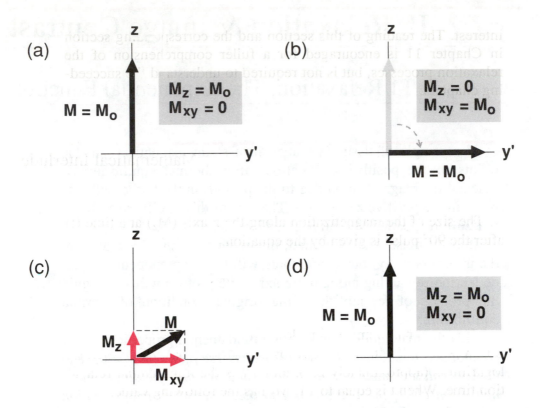

Fig. 8-1. Response of the magnetization to a 90° pulse and its subsequent relaxation, shown in the rotating frame. (a) At thermal equilibrium, the magnetization vector (M) points in the same direction as the static magnetic field (selected here, and throughout this book, as the positive z direction). The magnitude of the magnetization at thermal equilibrium is called M_0. (b) The application of a 90° pulse rotates M into the x-y plane, where it still has the size M_0. (c) After the 90° pulse, the reestablishment of thermal equilibrium begins: in this process M_z changes from 0 to M_0, and M_{xy} changes from M_0 to 0. (d) Reestablishment of the original condition of thermal equilibrium is complete.

magnetization along the z-axis, while a long T1 means a slow restoration. Sometimes the longitudinal relaxation is described using a relaxation rate (1/T1) instead of a relaxation time. A slow relaxation rate (1/T1) means a long relaxation time (T1); in other words, a small 1/T1 means a large T1. The T1 relaxation rate or relaxation time in a particular tissue depends on the properties of the tissue.

An understanding of the exponential function describing T1 relaxation is not necessary to appreciate the importance of T1. However, some general knowledge of exponential functions is helpful, and may make the reading of articles containing such functions less discouraging and more enlightening. The remaining part of this chapter is meant for the reader who has some mathematical interest. The reading of this section and the corresponding section in Chapter 11 is encouraged for a fuller comprehension of the relaxation processes, but is not required to understand the succeeding chapters.

Mathematical Interlude

The size of the magnetization along the z-axis (M_z) at a time (t) after the 90° pulse is given by the equation:

$$M_z = M_0 (1 - e^{-t/T1}) \qquad (8\text{-}1)$$

M_0 is the maximum attainable magnetization, i.e., the magnetization at complete thermal equilibrium; e is the base of the natural logarithms (approximately 2.7); and T1 is the longitudinal relaxation time. When t is equal to T1, M_z has the following value:

$$M_0\left(1 - e^{-1}\right) = M_0\left(1 - \frac{1}{2.7}\right) = 0.63\, M_0$$

In other words, after a time equal to T1, 63% of the maximum magnetization is regained. Eq. 8-1 describes the T1 relaxation curve shown in Fig. 8-2. It is easier to understand that this is an exponential curve if Eq. 8-1 is rearranged:

$$M_0 - M_z = M_0\, e^{-t/T1} \qquad (8\text{-}2)$$

Equation 8-2 now has the form $y = e^{-x}$. This basic exponential function is shown in Fig. 8-3. When the exponent (x) increases in the positive direction, y decreases towards zero. As x becomes increasingly negative, y increases toward infinity. When x is zero, y is 1, since $e^0 = 1$. In the exponential function $M_0 - M_z = M_0\, e^{-t/T1}$, t/T1 is the exponent corresponding to x. Since t is the time after the 90° pulse, t cannot have a negative value. Thus this particular exponential equation (Eq. 8-2) is valid only for values of $t/T \geq 0$,

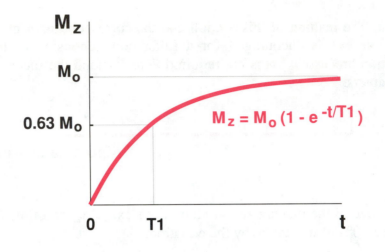

Fig. 8-2. Growth of the longitudinal magnetization (M_z) after the application of a 90° pulse; this is the T1 relaxation curve. M_0 is the magnitude of the magnetization vector at thermal equilibrium; t is the time after the 90° pulse. T1 is the time constant in the exponential function $M_z = M_0 (1 - e^{-t/T1})$, which describes the relaxation curve. In general terms, T1 describes the time it takes for M_z to relax toward M_0. At a time T1 after the 90° pulse, M_z equals 0.63 M_0, i.e., 63% of the longitudinal magnetization has been restored.

and its corresponding relaxation curve can be drawn as in Fig. 8-4. With increasing time (t) after the 90° pulse, $(M_0 - M_z)$ decreases towards zero, i.e., the magnetization along the z-axis (M_z) approaches M_0. Immediately after the 90° pulse, i.e., when t is zero, $e^{-t/T1} = e^0 = 1$; therefore, $M_0 - M_z = M_0$, meaning that M_z is zero.

In the literature, it is common to see the T1 relaxation curve drawn in a semi-logarithmic diagram, i.e., with logarithmic values along the ordinate (y-axis) and time along the abscissa (x-axis). To understand how this is done, we need to know two simple rules:

1) $\ln e^x = x$
2) $\ln (a \cdot b) = \ln a + \ln b$

Rule 1 says that the natural logarithm of e^x is equal to x; rule 2 says that the natural logarithm of a product is equal to the sum of the natural logarithms of each of the factors. We may even add a third, rather obvious "rule": If a = b, then ln a = ln b. Using this last statement, Eq. 8-2 may be written as:

$$\ln (M_0 - M_z) = \ln (M_0 e^{-t/T1}) \qquad (8\text{-}3)$$

Using rule number one and two above, Eq. 8-3 may be transformed into:

$$\ln(M_0 - M_z) = \ln M_0 + \ln e^{-t/T1} = \ln M_0 + \left(-\frac{t}{T1}\right) \qquad (8\text{-}4)$$

i.e.,

$$\ln(M_0 - M_z) = -\frac{1}{T1}t + \ln M_0 \qquad (8\text{-}5)$$

Eq. 8-5 now has the form of the linear function $y = ax + b$. If the values of $\ln(M_0 - M_z)$ are plotted along the y axis, and t along the x-axis, Eq. 8-5 will yield a straight line with the slope $a = -1/T1$ (Fig. 8-5). 1/T1 is called the **T1 (or longitudinal) relaxation rate**. A short T1 means a high relaxation rate and thus a steep negative slope for the line in Fig. 8-5; this means that $\ln(M_0 - M_z)$ decreases rapidly and that M_z approaches M_0 quickly. The relaxation rate is a useful concept, particularly when considering the effect of para-magnetic substances on the relaxation process (Chapter 16).

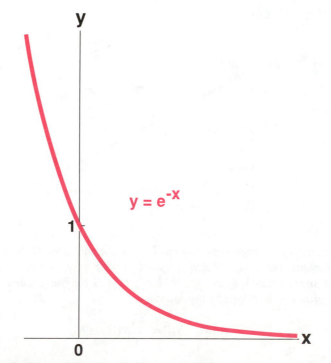

$$y = e^{-x}$$

Fig. 8-3. The exponential function $y = e^{-x}$. When $x = 0$, $y = 1$; as x becomes large, y approaches 0.

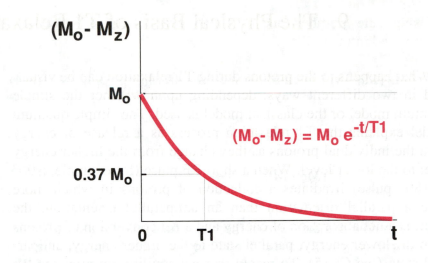

Fig. 8-4. *An alternate way of graphing the same T1 relaxation phenomenon as Fig. 8-2. Here, $M_0 - M_z$ is plotted along the z-axis instead of M_z. $M_0 - M_z$ is the difference between the actual longitudinal magnetization and its equilibrium value; thus, this value must approach zero as time elapses. This exponential function is described by the following equation:* $(M_0 - M_z) = M_0 e^{-t/T1}$.

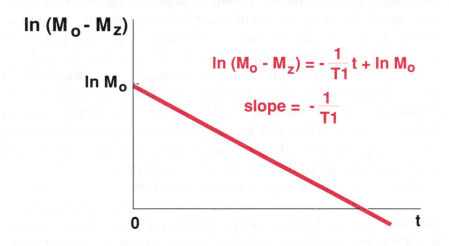

Fig. 8-5. *One more way of graphing the T1 relaxation curve. Here $\ln (M_0 - M_z)$ is plotted along the z-axis. When graphed in this way, the relaxation curve becomes a straight line whose slope is $-1/T1$; $1/T1$ is the longitudinal relaxation rate. This graph is described by the equation:*

$$\ln (M_0 - M_z) = -\frac{1}{T1} t + \ln M_0$$

9. The Physical Basis of T1 Relaxation

What happens to the protons during T1 relaxation can be visualized in two different ways, depending upon whether the simple quantum model or the classical model is used. The simple quantum model explains the T1 relaxation process as a release of energy from the individual protons as they change from the higher energy level to the lower level. When a short RF pulse (for example, a 90° or 180° pulse) irradiates a collection of protons in which more have a parallel orientation than an antiparallel orientation, the result is a net absorption of energy and a net conversion of protons from the lower energy, parallel state to the higher energy, antiparallel state (see Ch. 5). To regain thermal equilibrium after the RF pulse, there must be a net conversion of protons from antiparallel to parallel and hence a net release of energy.

A very small fraction of the energy that is released from the antiparallel protons is picked up by an antenna (the receiver coil), thus giving rise to the MR signal. Though it is not obvious from the simple quantum model, no signal is picked up by the receiver coil unless there exists some magnetization in the x-y plane. Nearly all of the energy released from the antiparallel protons is absorbed by the surrounding tissue rather than the receiver coil.

Generally, it is more productive to use the classical model (Ch. 7) to discuss the production of the signal induced in the receiver coil. Classically one can think of the MR signal as an electric current induced in the receiver coil by the precessing magnetization in the transverse plane. This electric current will be induced *only* if there exists a component of the magnetization in the transverse plane.

According to the simple quantum model, the rate of T1 relaxation is a measure of how fast the surplus number of antiparallel protons can get rid of the energy given to them by the photons in the RF pulse. Since the value of T1 differs for different tissues, there must be something in the tissues — in the chemical and physical environment — that affects the ease with which the antiparallel protons can get back to the parallel orientation. To explain this "something" that affects T1 relaxation we will first turn to the classical model.

As previously discussed, an RF pulse at the Larmor frequency will create a magnetic field vector (B_1) rotating at the Larmor frequency in the transverse plane. This rotating magnetic field will force the magnetization vector to spiral down into the transverse plane, further down into the antiparallel direction, back again to the parallel direction and so on, depending upon the duration of the RF pulse (Ch. 6). As the magnetization changes toward the antiparallel position, the protons experience a net absorption of energy; as the magnetization changes toward the parallel direction, the protons experience a net loss of energy. Thus we see that a radio wave at the Larmor frequency can cause the protons to both gain and lose energy.

Actually, any magnetic field fluctuating at the Larmor frequency can produce this same effect. **Magnetic "noise,"** always present in the body, will include some fluctuations of the magnetic field at the Larmor frequency. (This "noise" comes from sources within the tissues; it does not refer to RF noise that might come from outside sources.) These fluctuations lead to relaxation, by the same mechanism as explained for the magnetic field of the RF pulse, i.e., by creating a torque that changes the orientation of the proton with respect to the static magnetic field direction. Such fluctuating magnetic fields are present in the tissues of the body even in the absence of an RF pulse.

In addition to the strong, static magnetic field created by the MR magnet, all the protons in the body will also experience small, local magnetic fields produced by neighboring protons, or by other nuclei or molecules having a magnetic moment. Due to the thermal motions of the molecules, these local magnetic fields will fluctuate, i.e., produce magnetic "noise." These fluctuations are multi-directional and consist of many frequencies. Those components of the local magnetic fields having the Larmor frequency and an appropriate orientation can rotate the protons from an antiparallel to a parallel orientation, and vice versa.

The orientation of a proton with respect to the static field direction is not fixed. Even at thermal equilibrium, the protons constantly change their orientations. This changing orientation of the protons is a prerequisite for T1 relaxation to occur.

Let us return to the simple quantum model to see how magnetic "noise" at the Larmor frequency can change the relative number of parallel and antiparallel protons. In Chapter 5 we saw that when the photons in a radio wave cause the protons to change between their two allowed orientations, the probability of changing a parallel proton into an antiparallel proton was the same as the probability of changing an antiparallel proton into a parallel one. If there were more parallel than antiparallel protons then there would be a net increase in the numbers of antiparallel protons. If the transitions between parallel and antiparallel orientations caused by the magnetic "noise" in the tissue obeyed the same rule of equal probabilities, the resulting condition of thermal equilibrium would be equal numbers of parallel and antiparallel protons, since there would no preference for either orientation.

This, of course, is not the case. To obtain the equilibrium condition — a slightly greater number of parallel protons as compared to antiparallel protons — we must have the following rule for transitions caused by magnetic "noise": the probability of changing an antiparallel proton into a parallel proton is slightly greater than the probability of changing from parallel to antiparallel. In fact, the ratio of the first to second probability is the same as the ratio of the number of parallel to antiparallel protons at thermal equilibrium. This is a result of the proton's preference to exist in its lower energy state corresponding to parallel orientation. As T1 relaxation occurs, the net change of protons from antiparallel to parallel alignment releases energy which is transferred to the thermal energy of the random molecular motions.

In the language of the quantum model: After an RF pulse, the reestablishment of the magnetization along the static magnetic field direction is due to the preference for the lower energy state of the parallel protons. However, this preference could not alone change the magnetization without the fluctuating magnetic fields, which produce a continual transformation of the protons between parallel and antiparallel orientations. If the parallel and antiparallel protons just sat there without flipping, a change in their relative numbers could not occur. *The strength of the T1 relaxation process in a particular tissue depends on the strength of the magnetic field fluctuations or "noise" in that tissue experienced by the protons and occurring at the Larmor frequency. As these fluctuations, or "noise," become stronger, T1 becomes shorter.*

Just as audio noise is made up of a range of sound frequencies, the magnetic "noise" in tissues is made up of magnetic field fluctuations occurring over a range of frequencies. The frequencies comprising the magnetic "noise" correspond to the frequencies of the molecular thermal motions in the tissue. (These molecular thermal motions include linear motion, rotation, and vibration.) Slowly moving molecules have only low motional frequencies. Their contribution to the magnetic "noise" is therefore primarily at low

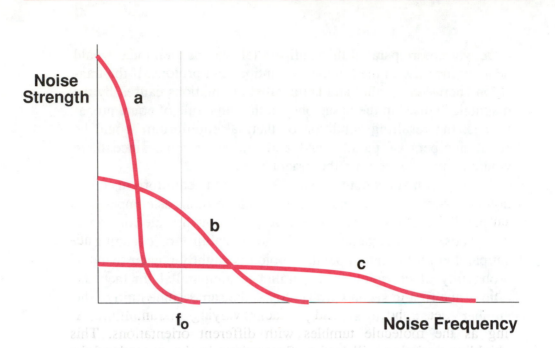

Fig. 9-1. The strength of the magnetic "noise" due to nuclear magnetic moments in the tissue. This graph shows the amount of magnetic "noise" existing at different frequencies; f_0 is the Larmor frequency. All noise at frequencies above DC (or zero frequency, at the extreme left of the graph) are due to the thermal motions of the magnetic moments. Curve a corresponds to the "noise" found in solids which only have very slow molecular motions; it also shows the noise due to macromolecules (in a fluid) that move very slowly due to their large size. Curve b corresponds to viscous fluids or to moderately sized molecules, such as lipids, in a fluid. Curve c corresponds to the fastest moving, small molecules: freely moving water in a non-viscous fluid.

frequencies. Rapidly moving molecules contribute a much wider range of frequencies to the magnetic "noise." This is shown in Fig. 9-1. As molecular motion increases, its contribution to low frequency magnetic "noise" decreases while the contribution to high frequency magnetic "noise" increases. The total magnetic "noise" present at each point in the tissue is the result of the contributions from neighboring molecules.

Actually, to properly explain the relaxation process we must consider the magnetic "noise," not at a fixed point in space, but at the positions of the protons which are moving. Therefore, the magnetic "noise" affecting the relaxation of a proton will be the result of the motion of both the proton and its neighbors. The most important neighbor to the proton in a water molecule is the other proton (or hydrogen nucleus) in the molecule. The thermal motion affecting both the original proton and its most important neighbor is simply the tumbling motion of the water molecule in which they both reside.

Thus, the main source of the magnetic "noise" causing T1 relaxation is usually the magnetic moments of the protons themselves, and for any specific proton in a water molecule, the principle proton contributor is the other proton in the molecule. Two additional minor sources should, however, be mentioned.

The protons are parts of molecules and are therefore surrounded by "clouds" of electrons. In such a "cloud" of electrons an electric current will be induced by the strong static magnetic field. This electric current will create a magnetic moment vector in the anti-parallel direction, which will reduce the local magnetic field strength experienced by the proton. This effect is called **electronic shielding**. The distribution of electrons surrounding a proton can be asymmetric, however, and produce a varying amount of shielding as the molecule tumbles with different orientations. This shielding variation will lead to fluctuations in the strength of the local magnetic field experienced by the proton, and if these fluctuations contain frequencies close to the Larmor frequency of the proton, T1 relaxation will be promoted. Yet another source of fluctuating magnetic fields is the presence of paramagnetic molecules in the tissue. The mechanisms by which paramagnetic substances enhance T1 (and T2) relaxation will be discussed in Chapter 16.

The range of frequencies of a molecule's thermal motions depends upon the size of the molecule and upon whether the molecule's motion is impeded by other molecules due to the structure of the substance. In a fluid, large molecules such as proteins will move more slowly than small molecules such as water. As a fluid becomes more viscous all molecules are slowed down. In a solid, the molecular motions are much slower than in a fluid. The local magnetic field fluctuations in solids are composed mostly of frequency components at very low frequencies, well below the Larmor frequency of the protons (curve *a* in Fig. 9-1). Protons in solids therefore have a very slow T1 relaxation and hence a very long T1 (Fig. 9-2).

Most tissues in the body, other than bone and cartilage, are fluids composed principally of water molecules with the addition of other molecules of greatly varying size, including larger molecules such as lipids and proteins. These larger molecules also contain protons. In a solution of pure water or in tissues with a high proportion of fluid water, the thermal motions of the water molecules will be relatively unimpeded and will correspond to the fast moving molecules shown in curve c of Fig. 9-1. The protons in such a fluid will experience magnetic "noise" over a wide range of frequencies from zero to well above the Larmor frequency. Only a small amount of this noise will occur at the Larmor frequency and thus this magnetic "noise" is not very good at producing T1 relaxation. This means that T1 will be relatively long (Fig. 9-2). CSF is an example of such a fluid with a T1 that is long (approximately 2 to 4 seconds) (25, 29, 33, 58, 64) compared to the T1 in most soft tissues of the body (approximately 0.1-1.0 seconds) (25, 33, 44, 58, 64).

If large, water-attracting molecules like proteins are added to this fluid, the motions of the water molecules will be affected. Many of these water molecules will attach to the protein or be influenced by it so that their thermal motions are slowed. A significant fraction of the water molecules will frequently

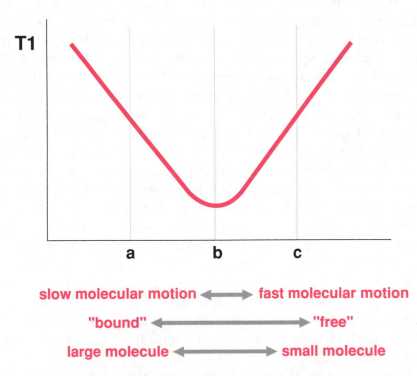

Fig. 9-2. The dependence of T1 on the speeds of the molecular thermal motions. The points a, b, and c on this graph correspond to curves a, b, and c respectively in Fig. 9-1. T1 relaxation is enhanced by molecular motions with moderate speeds, which therefore produce the shortest T1s. T1 lengthens with very slow or very fast molecular motions. Freely moving molecules have faster molecular motions than those that are bound; also, smaller molecules have faster molecular motions than larger molecules.

exchange between a "free" state and relatively "bound" states near the proteins. The average magnetic "noise" experienced by these free, bound, and exchanging protons is better represented by curve *b* in Fig. 9-1. The range of noise frequencies will still include the Larmor frequency, but since the frequency range does not now go as high as in curve *c*, a greater proportion of the "noise" exists at Larmor frequency. This produces enhanced T1 relaxation and a shorter T1 (Fig. 9-2).

If moderately sized molecules such as lipids are the principal constitutents of the fluid, the larger size of these molecules relative to water makes their thermal motion inherently slower than free water. Like the protons in water that are slowed down by proteins, the magnetic "noise" experienced by the protons in the lipids will also be better represented by curve *b* in Fig. 9-1. This means that the protons in fat will also have a shortened T1, compared to pure water (Fig. 9-2).

Some lipids are bound into membranes and are relatively non-mobile. Proteins are also relatively non-mobile due to their large size. The protons in these substances will experience magnetic "noise" more like that in curve *a* in Fig. 9-1. Relatively non-mobile protons, like those in proteins and in non-mobile lipid chains in membranes, do not contribute to the MR image for reasons we shall discuss in Chapter 12.

Note that the resultant T1 which is measured in a region of interest in the tissue is actually the net result of protons undergoing many different relaxation processes. Some of the protons in that region may be located in freely moving water molecules, while others are in water molecules attached to or associated with macromolecules such as proteins. Some of the protons may be found in fat molecules. The T1 relaxation of the protons in these different environments will all occur at different rates. Depending upon the composition of the tissue, these different relaxation rates will combine to give the characteristic T1 value of that particular tissue.

In summation, T1 (longitudinal) relaxation is due to an interaction of the magnetic moments of the protons with local fluctuating magnetic fields (magnetic "noise") in the tissue. The magnetic "noise" seen by a proton depends on its random thermal motions and the motions of its close neighbors in the tissue. The thermal motions of a molecule are determined by the size of the molecule and whether it is influenced — and slowed down — by neighboring molecules (such as proteins). The T1 relaxation process entails a net transfer of energy from the proton magnetic dipoles to the thermal energy in the tissue. This process of energy transfer is reflected in the alternative term for this process: **spin-lattice relaxation**. "Spin" is used here as just another name for the proton. "Lattice" originally referred to the molecular environment of the proton in solids having a lattice structure; in today's use of the term its reference is extended to any physical environment of the proton, including living tissue.

10. Image Contrast Due to T1 Relaxation, Saturation-Recovery

In Chapter 7 we saw that immediately after a 90° pulse is applied to tissue at thermal equilibrium in a static magnetic field, a signal called the FID (free induction decay) is produced in the receiver coil. In this process the equilibrium magnetization, M_0, which originally points along the static magnetic field direction, is rotated into the transverse plane so that at the start of the FID the magnetization in the transverse plane (M_{xy}) has the magnitude M_0. As time progresses the transverse component of the magnetization (M_{xy}) decreases from M_0 toward zero, while the longitudinal component (M_z) grows from zero toward the value M_0. If another 90° pulse is applied to the tissue *before* the longitudinal component has returned to its equilibrium value of M_0, the magnetization that is rotated into the transverse plane will be less than M_0 and the amplitude of the resulting FID signal will be reduced compared to the first signal (Fig. 10-1).

The amplitudes of both FID signals are affected by the magnitude of the equilibrium magnetization, M_0, which in turn is affected by the proton density (number of protons per unit mass) in the tissue. (We only include the protons that can contribute to the FID signal. As noted previously, this includes the protons in "mobile" water and lipids but not the relatively "non-mobile" protons in proteins and membrane lipids.) The amplitude of the latter FID is also affected by the following properties: (1) the T1 of the tissue, and (2) the time interval between the 90° pulses. A longer T1 or a shorter time interval will result in a smaller FID signal.

The simplest way to obtain image contrast dependent upon T1 is to transmit a series of successive 90° pulses (Figs. 10-2 and 10-3).

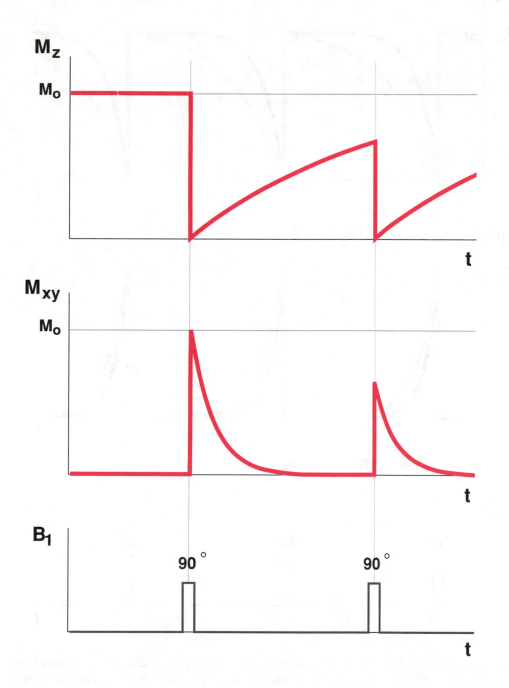

Fig. 10-1. Application of two successive 90° pulses after the tissue is at thermal equilibrium. Immediately after the first 90° pulse, M_{xy} is equal to the equilibrium magnetization M_0. If a second 90° pulse is applied before complete T1 relaxation has occurred, M_{xy} after this 90° pulse will be reduced in magnitude. In such a case, the size of the second FID will be less than that of the first FID. (The size of the FID is proportional to the size of M_{xy}.)

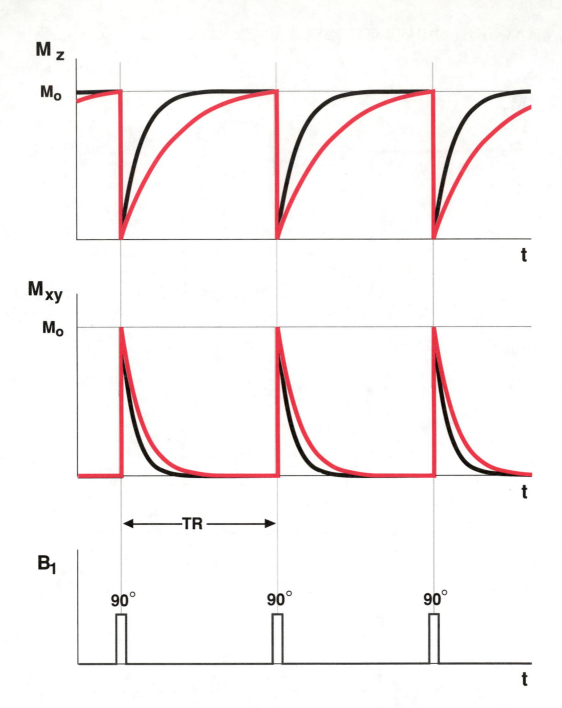

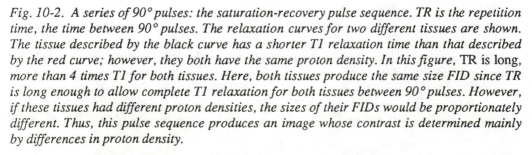

Fig. 10-2. A series of 90° pulses: the saturation-recovery pulse sequence. TR is the repetition time, the time between 90° pulses. The relaxation curves for two different tissues are shown. The tissue described by the black curve has a shorter T1 relaxation time than that described by the red curve; however, they both have the same proton density. In this figure, TR is long, more than 4 times T1 for both tissues. Here, both tissues produce the same size FID since TR is long enough to allow complete T1 relaxation for both tissues between 90° pulses. However, if these tissues had different proton densities, the sizes of their FIDs would be proportionately different. Thus, this pulse sequence produces an image whose contrast is determined mainly by differences in proton density.

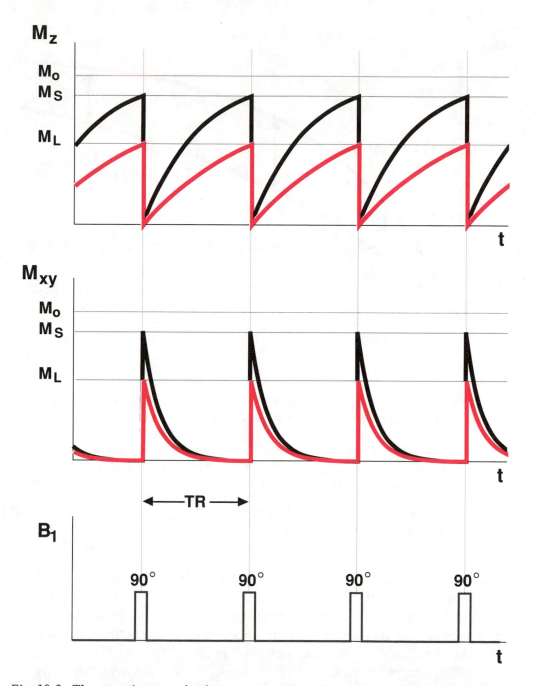

Fig. 10-3. The same tissues and pulse sequence as Fig. 10-2 except that TR has been reduced so that it is comparable to the T1 relaxation times of the tissues. Full T1 relaxation does not occur between 90° pulses. Here, the tissue with the longer T1 relaxation time (red) produces the smaller FID signal. If these tissues had different proton densities the sizes of their FIDs would be also be affected by the proton density. Thus, this pulse sequence would produce an image whose contrast is determined by differences in both T1 relaxation and proton density.

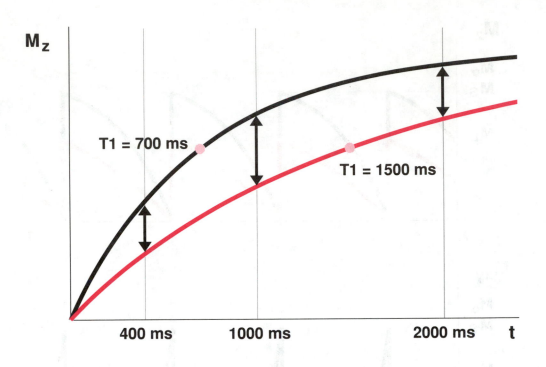

Fig. 10-4. Optimizing image contrast from T1 relaxation. This figure shows the T1 relaxation curves of two tissues, one with T1 = 700 ms (black), the other with T1 = 1500 ms (red). The separation between the curves at any point in time gives the contrast produced by a saturation-recovery pulse sequence whose TR is equal to that time (ignoring variations in proton density). The greatest contrast occurs at a TR that is between the T1 values of the tissues. In this case a TR of 1000 ms would yield the greatest contrast; a TR of 400 ms or 2000 ms would result in 25% less contrast. (For those interested, the value of TR yielding the greatest contrast between two tissues is $TR_{optimum} = ln\ (T1_A/T1_B)\ /\ [(1/T1_B) - (1/T1_A)]$. Here, $T1_A$ and $T2_B$ are the relaxation times of the two tissues, with $T1_A > T1_B$.)

The time interval between each 90° pulse is called the **repetition time (TR)**. If the repetition time is at least 4–5 times the longest T1 in the slice of tissue being imaged, then all the tissue components in the slice will have fully regained their magnetization along the z-axis (M_z) when the next 90° pulse is applied (Fig. 10-2). FIDs, derived from the different voxels of the slice, will then only reflect differences in proton density. When the repetition time is shorter, however, the magnetization vectors in the different voxels will not have reached their maximum size when the next 90° pulse is transmitted, and the resulting induced FIDs will have a reduced amplitude (Fig. 10-3). Among tissues with the same proton density, *the weakest FID signal will be derived from the tissue having* *the longest T1, and this tissue will therefore appear dark in the image. The brightest areas in the image will correspond to the tissue with the shortest T1. In actual tissues, the contrast in the image is determined partly by differences in proton density, partly by differences in T1.*

Maximum T1 contrast between two tissues is achieved when the repetition time is close to the average T1 of the two tissues. Such a TR provides the largest separation of the T1 relaxation curves at the time of RF pulse transmission (Fig. 10-4).

The pulse sequence shown in Figs. 10-2 and 10-3 is called **saturation-recovery (SR)**. The term "saturation" refers to the fact that immediately after a 90° pulse, there is no magnetization along the z-axis. A new 90° pulse transmitted at that time would result in no FID signal, as there would be no longitudinal magnetization to rotate into the transverse plane. The result would, instead, be a rotation of the transverse magnetization into the antiparallel direction. Therefore, immediately after a 90° pulse, the protons in the tissue are said to be **saturated**. Immediately before the next 90° pulse the longitudinal magnetization has "recovered" by varying degrees in different parts of the tissues giving rise to the tissue contrast.

The term **partial-saturation (PS)** is, at times, used interchangeably with saturation-recovery. Partial-saturation refers to the fact that immediately before each of the 90° pulses the transverse magnetization is neither zero (fully saturated) nor equal to M_0 (fully recovered) — rather, it is partially saturated. (This is just a different way of expressing what was said a few sentences previously.) Often, however, partial saturation is used as a synonym for the spin-echo pulse sequence with short TR and short TE (echo time) (Ch. 14).

11. T2 Relaxation, T2 vs. T2*

So far, we have seen how the contrast in MR images can be determined by two parameters, the proton density and T1, if the saturation-recovery pulse sequence is used. With the use of other types of pulse sequences, a third parameter can also have a very significant effect on image contrast. This parameter is T2, the time constant of the process called **T2 (or transverse) relaxation**.

As discussed previously, the reestablishment of thermal equilibrium after a 90° pulse entails both a regaining of the magnetization in the static magnetic field direction (T1 relaxation) and a loss of the magnetization in the transverse plane. Some of the loss of transverse magnetization evidently must be related to the processes that produce T1 relaxation. As the magnetization along the static magnetic field direction is restored, the component of the magnetization in the transverse plane must decrease. (When it is fully restored there can be no transverse magnetization remaining.)

The T1 relaxation process is relatively slow, however, and usually the magnetization in the transverse plane has disappeared long before the T1 relaxation is completed. The reason for this fast loss of transverse magnetization is the dephasing of the protons which comprise the spin excess.

In Chapter 7 we saw that the magnetic moments of the protons comprising the spin excess precessed together (in phase) immediately after a single 90° pulse; in other words, these magnetic moments all pointed in the same direction as they precessed in the transverse plane. Thus, at that instant, their magnetic moments added together to give the maximum possible value for the transverse magnetization: the transverse magnetization (M_{xy}) is here equal in size to that of the equilibrium value of the longitudinal magnetization (M_0). As time passes, the proton magnetic moments dephase, that is, they precess with different frequencies and thus do not all point together in the same direction (Fig. 11-1). The dephasing of these magnetic moments is the reason for the quick loss of the transverse magnetization (M_{xy}) and the resultant decay of the FID signal.

M_{xy} and the FID signal decay approximately exponentially; **T2*** is the time constant of the exponential function that approximately describes this decay (Fig. 11-2). After a time equal to T2*, the amplitude of the FID (i.e., the signal intensity) is reduced to 37% of its maximum value.** After a time equal to 4 to 5 times T2* there is essentially no remaining FID signal.

The key to understanding the dephasing of the protons is the Larmor equation, $f = (\gamma/2\pi)B$, which states that the precessional

---

** The 37% of M_{xy} that remains after a time T2* should be compared to the 63% of M_z that is regained after a time T1 (Chap 8). (Note that 37% + 63% = 100%.) In the relaxation processes so far described, M_{xy} decreases with time (T2* relaxation) while M_z increases with time (T1 relaxation). In both cases, 63% of the total change in M_{xy} or M_z has occurred after a time T2* or T1, respectively. Since 63% of M_{xy} has disappeared after a time T2*, 37% must be left.

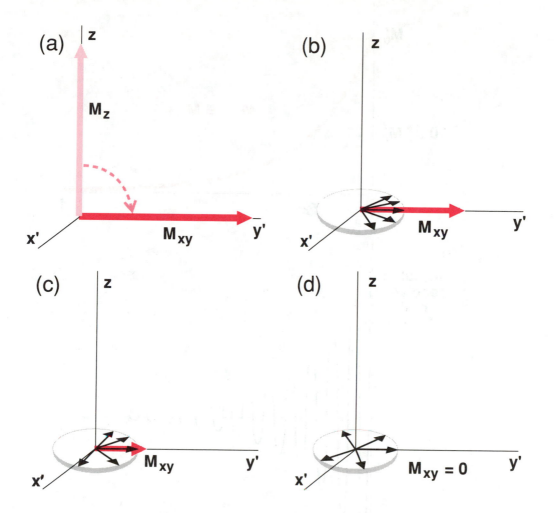

Fig. 11-1. Dephasing of the proton magnetic moments causing T2* decay of the transverse magnetization, depicted in the rotating frame. (a) The application of a 90° pulse rotates the magnetization from the z-axis (the direction of the static magnetic field) to the x-y plane, without changing its size. (b) and (c) As time passes, the individual proton magnetic moments dephase and M_{xy} decreases. This figure shows the magnetic moments of five representative protons experiencing slightly different magnetic fields. (d) Complete dephasing of the proton magnetic moments results in $M_{xy} = 0$. In (b), (c), and (d) T1 relaxation also causes a progressive increase in M_z from zero, though this is not shown in the figure.

frequency of the proton is proportional to the strength of the static magnetic field. Proton dephasing means that the magnetic moment vectors of different protons are precessing at slightly different frequencies. This is caused by slight differences in the local magnetic field strength experienced by the individual protons. Such variations in the strength of the local magnetic fields are caused

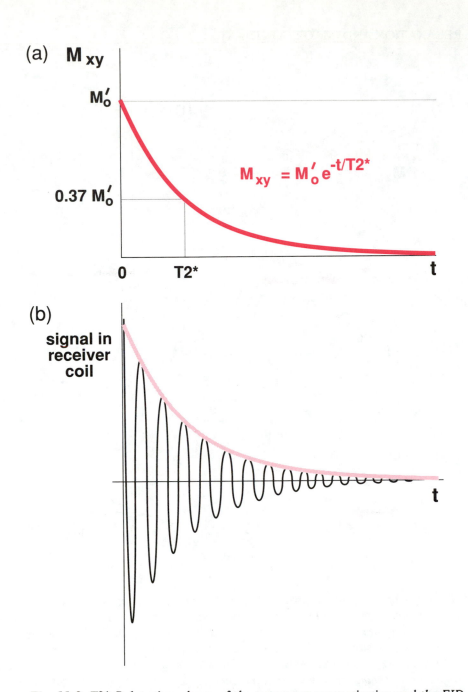

Fig. 11-2. T2* Relaxation: decay of the transverse magnetization and the FID. (a) M'_0 is the magnitude of the transverse magnetization immediately after the 90° pulse; t is the time after the 90° pulse. (Note: If complete T1 relaxation has occurred before the 90° pulse is applied, then M'_0 will equal M_0; if complete T1 relaxation has not occurred at this time, then M'_0 will be less than M_0.) After the application of a 90° pulse, the transverse magnetization (M_{xy}) decreases with time in a roughly exponential manner: $M_{xy} = M'_0 e^{-t/T2*}$. The time constant of this decay is called T2*. After a time T2*, about 37% of the transverse magnetization remains. (b) Here, the black oscillating curve represents the signal in the reciever coil, the FID. The amplitude or envelope of the FID, shown in red, is the same curve as the decay curve in (a).

by: (1) variations produced by the MR magnet, and (2) variations produced by chemical/physical processes in the tissue itself.

The first case is due to constant inhomogeneities in the strong, static magnetic field of the MR magnet. Although MRI is dependent upon a very homogeneous static magnetic field, no MR magnet is perfect. Even within a volume of tissue as small as an imaging voxel, the static field varies slightly from one location to the other, thus causing a dephasing of the protons within the voxel.

Variations in the magnetic field in the second case are due to the inherent properties of the tissue being imaged and are independent of the inhomogeneities in the static field produced by the MR magnet. Even if the static magnetic field of the magnet was perfectly homogeneous, the protons would still dephase due to random variations in the local magnetic field strength created by the physical and chemical environment of the proton. These magnetic field variations are part of the same magnetic "noise" that was discussed in explaining T1 relaxation. Proton dephasing, however, is produced by the low frequency part of this noise, in other words by magnetic field fluctuations that are relatively slow (see Chap. 12). This tissue-dependent loss of transverse magnetization is called **T2** (or **transverse) relaxation**; T2 is the time constant of the exponential function that describes this component of the decay of the transverse magnetization.

In describing tissue properties we are concerned with T2, *not* T2*. Since T2* is affected by the degree of inhomogeneity of the MR magnet, its value for a particular tissue will vary when measured with different MR equipment. (In Chap. 13, we will describe a technique that can eliminate the effect of magnet inhomogeneities. This will allow us to measure T2 directly rather than T2* even in the presence of magnet inhomogeneities.)

Imagine an idealization in which we place tissue in a perfectly homogeneous static magnetic field and apply a 90° pulse. In this case we could measure T2 directly by observing the FID signal. The amplitude of this signal would be proportional to M_{xy}. By definition, a plot of M_{xy} versus time is the T2 relaxation curve if the static magnetic field is perfectly homogeneous. (In this case, T2* would be equal to T2.) At a time T2 after the 90° pulse, the magnetization in the transverse plane (and thus the signal in the receiver coil) would be reduced to 37% of its original value due to T2 (or transverse) relaxation (curve B in Fig. 11-3a). (Or, to put it

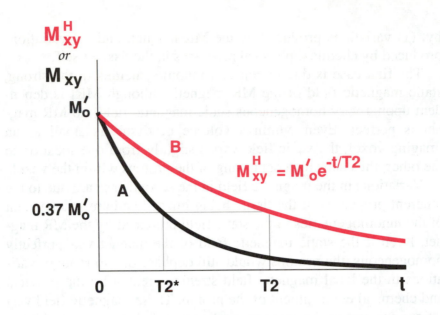

Fig. 11-3. *T2 versus T2* relaxation. Curve A is the same curve as in Fig. 11-2 (a). It shows the actual decay of the transverse magnetization, M_{xy}, with a time constant of T2*. Curve B describes an idealized situation in which the static magnetic field produced by the MR magnet is perfectly homogeneous. In this case, the decay of the transverse magnetization is caused only by the magnetic field inhomogeneities produced by the magnetic moments in the tissue itself. This idealized decay is described by an exponential function with a time constant of T2: $M_{xy}^{H} = M'_{0}e^{-t/T2}$. We can talk about the T2 decay even when using a real, imperfect magnet; to distinguish this idealized decay from the actual decay of M_{xy}, we introduce the term M_{xy}^{H}. M_{xy}^{H} is the value that the transverse magnetization would have if the static magnetic field of the MR magnet were perfectly homogeneous. After a time T2, M_{xy}^{H} is reduced to 37 percent of its original value. While the M_{xy} decay curve (and thus the value T2*) is affected by the characteristics of the MR magnet, the M_{xy}^{H} decay curve (and thus the value T2) is determined only by the characteristics of the tissue itself and not by any imperfections in the magnetic field produced by the MR magnet. T2* is always less than T2.*

another way, 63% of the original transverse magnetization would have decayed away.)

A short T2 thus means a rapid loss of transverse magnetization, while a long T2 means a slow loss of this vector. In reality, by measuring the decay time of the FID after a 90° pulse we are actually determining T2* rather than T2 since no magnet is perfect (curve A in Fig. 11-3). T2* is always shorter than T2; in fact, it can be much shorter if T2 is long and the magnet is not of high quality. We shall see in Chap. 13 that there are other pulse sequences using the phenomenon of "echoes" that allow us to effectively compensate for the imperfections in the magnet and to come close to a true measurement of T2.

At certain points in the remainder of this book it will be convenient to talk about the size that the transverse magnetization would have if the static magnetic field produced by the MR magnet were perfectly homogeneous. We will use the term M_{xy}^H to refer to this idealized transverse magnetization. M_{xy}^H is not a specific term that you will find in the rest of the MR literature, but it is a useful tool for our explanations of MR processes. In practice, the actual transverse magnetization, M_{xy}, will always be less than M_{xy}^H once the decay of the transverse magnetization begins. By definition, the T2 relaxation curve is the plot of M_{xy}^H versus time.

The remaining part of this chapter is meant for the mathematically interested reader. Knowledge of its contents is not required to understand T2 relaxation at a basic level.

Mathematical Interlude

The exponential function describing T2 (transverse) relaxation is:

$$M_{xy}^H = M'_0 \, e^{-t/T2} \qquad (11\text{-}1)$$

In this equation, M'_0 is the maximum magnetization in the transverse plane; it is the size of the transverse magnetization immediately after the 90° pulse. M_{xy}^H is the size of the transverse component of the magnetization vector at a time (t) after the 90° pulse (assuming a perfectly homogeneous static magnetic field), and T2 is the transverse relaxation time constant. When t is equal to T2, M_{xy}^H becomes equal to:

$$M'_0 \, e^{-1} = \frac{1}{e} M'_0 = \frac{1}{2.7} M'_0 = 0.37 \, M'_0 \qquad (11\text{-}2)$$

In other words, the transverse magnetization is reduced to 37% of its maximum value after a time equal to T2. When t is zero, M_{xy}^H is equal to M'_0, since $e^0 = 1$. As t increases, $e^{-t/T2}$ decreases, and M_{xy}^H approaches zero.

Eq. 11-1 can be transformed by taking the logarithm of both sides of the equation:

$$\ln M_{xy}^{H} = \ln \left(M'_0 \, e^{-t/T2} \right) \tag{11-3}$$

$$\ln M_{xy}^{H} = -\frac{1}{T2} t + \ln M'_0 \tag{11-4}$$

The T2 relaxation equation is now expressed as a linear function $y = ax + b$, instead of as an exponential function. If the values of $\ln M_{xy}^{H}$ are plotted along the vertical axis, and t along the horizontal axis, Eq. 11-4 will yield a straight line with the slope $a = -1/T2$ (Fig. 11-4). $1/T2$ is called the **T2 relaxation rate**. A short T2 thus means a high relaxation rate, a fast decay of the transverse magnetization, and a steep negative slope for the line in Figure 11-4.

Similar to the exponential function of T2 relaxation, the actual decay of the transverse magnetization using a real magnet with inhomogeneities can be approximated by the equation:

$$M_{xy} = M'_0 \, e^{-t/T2*} \tag{11-5}$$

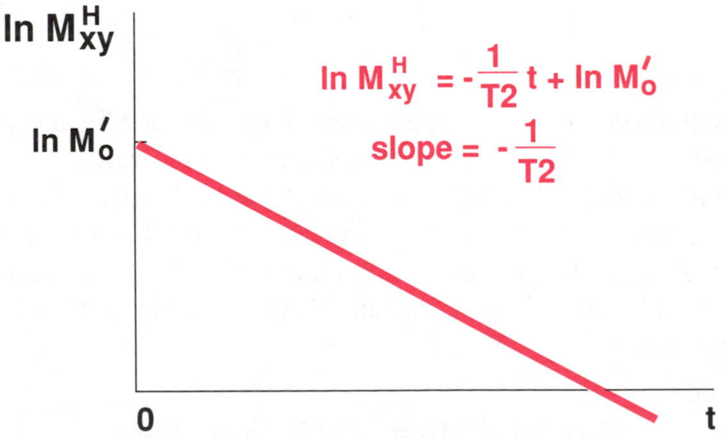

In the figure:

$$\ln M_{xy}^{H} = -\frac{1}{T2} t + \ln M'_0$$

$$\text{slope} = -\frac{1}{T2}$$

Fig. 11-4. Another way of graphing the T2 relaxation curve. Here $\ln M_{xy}^{H}$ is plotted along the vertical axis. When graphed in this way, the relaxation curve becomes a straight line whose slope is -1/T2; 1/T2 is the transverse relaxation rate. This graph is described by the equation:

$$\ln M_{xy}^{H} = -\frac{1}{T2} t + \ln M'_0$$

In this case, M_{xy} is the actual transverse magnetization at a time (t) after the 90° pulse, and M'_0 — as before — is the value of the transverse magnetization immediately after the 90° pulse. Since the FID signal is proportional to M_{xy}, this equation also gives the approximate shape of the FID. T2* is the time constant for the decay of the actual transverse magnetization (or of the FID signal). The decay rate is 1/T2*. The part of this decay that is due solely to field inhomogeneities produced by the MR magnet has a rate we will call $1/T2_{inh}$. The decay rate 1/T2* is the sum of the decay rate due to the static magnetic field inhomogeneities plus the decay rate due to the magnetic "noise" in the tissue, i.e., the T2 relaxation rate. Thus:

$$1/T2^* = 1/T2 + 1/T2_{inh} \qquad (11\text{-}6)$$

You should note that when two independent relaxation processes affect the same physical phenomenon (in this case, the loss of transverse magnetization), the net (observable) result is a relaxation rate equal to the sum of the individual *relaxation rates*. There is no such simple, additive relationship between the relaxation times themselves (i.e., T2*, T2 and $T2_{inh}$).

12. The Physical Basis of T2 Relaxation, Comparison of T2 & T1 Relaxation, Dependence of T1 & T2 on the Larmor Frequency

As mentioned in the previous chapters, both T1 and T2 relaxation are produced by randomly occurring variations in the local magnetic field strength, or magnetic "noise," in the tissues. We have discussed the sources of this "noise" and the manner in which molecular motions affect its frequency components. We have also seen that it is the part of the magnetic "noise" at the Larmor frequency that causes T1 relaxation. We shall see now that T2 relaxation is produced by the parts of this magnetic "noise" at both the low frequencies and the Larmor frequency.

In Chap. 11 we learned that T2 relaxation is due to the

dephasing of the individual magnetic moments of the protons comprising the spin excess. This dephasing begins immediately after the application of the 90° pulse. The protons will dephase if they each experience a slightly different magnetic field and thus precess at a slightly different frequency. We will, in this chapter, ignore the presence of inhomogeneities produced by the magnet (so that we are considering only the tissue characteristic, T2 effects and not T2*). Then the differing magnetic fields experienced by the protons will be due to the tissue's magnetic "noise."

The higher frequencies in this "noise" will cause the protons to experience a rapidly fluctuating change in the magnetic field. This will produce a corresponding fluctuation in the protons' precessional frequency. The "noise" at 1 Mhz, for example, will cause the protons' precession to speed up and then slow down one million times a second. Over the many milliseconds during which T2 relaxation occurs the effects of this "noise" on the phase of the protons will average out.

Thus the "noise" at high frequencies will not produce a net dephasing of the protons and will not contribute to T2 relaxation. (The exception is the "noise" at the Larmor frequency which we will discuss shortly.) To produce a net dephasing of the protons, the local magnetic field variations must be relatively constant over many milliseconds. This property applies to the part of the magnetic "noise" that is slowly varying — the part at low frequencies — thus this "noise" does contribute to T2 relaxation.

To see how T2 relaxation is affected by the magnetic "noise" at the Larmor frequency, we need to consider that T2 relaxation is in part related to the T1 relaxation process. When T1 relaxation is completed, by definition, the magnetization has reached its maximum value along the longitudinal direction and no magnetization remains in the transverse plane. This means that T1 relaxation, all by itself, forces T2 relaxation to occur; thus the magnetic "noise" at the Larmor frequency that contributes to T1 relaxation also contributes to T2 relaxation. There will also be additional contributions to T2 relaxation, as discussed above, by the low frequency magnetic noise. Because of these additional contributions, the T2 relaxation rate must always be larger than or equal to the T1 relaxation rate, or, in other words, *T2 can never be longer than T1*. In pure water the contributions to T2 relaxation from the "noise" at the Larmor frequency and at low frequency are essentially equal to the contributions to T1 at the Larmor frequency; this is the limiting case. Thus, in pure water T2 is nearly equal to T1. However, in almost every tissue in the body, T2 is considerably shorter than T1 (e.g., T2 = 50 msec and T1 = 500 msec).

Let us now look in more detail at how the magnetic "noise" at the Larmor frequency contributes to T2 relaxation. The proton dephasing process caused by low frequency "noise" does not involve a change in the energy of the proton magnetic dipoles, since it does not rotate them into greater or lesser alignment with the static magnetic field — it only affects their precession in the x-y plane. However, changes in the energy of the proton, produced by local magnetic field variations at the Larmor frequency, also contribute to proton dephasing and to T2 relaxation.

Using the simple quantum description, it is possible for a proton to change from the higher energy level to the lower level and give up energy in the process. This energy can be transferred to the thermal energy of the molecular motions of the tissue. This, we saw, is a T1 relaxation process. Another possible event is the absorption of this energy by another proton which then changes from the lower energy level to the higher level. This latter event is not a T1 relaxation process since the net number of parallel and antiparallel protons has not changed. However, in either event as a proton changes its state, it loses track of its phase relationship with the other protons. Thus both of these processes contribute to proton dephasing and to T2 relaxation.

When the proton changes energy states it will change the direction of the small magnetic field that it produces; thus the local magnetic field felt by neighboring protons will change. This change contributes to the local magnetic field fluctuations and promotes both T1 relaxation and T2 relaxation.

The exchange of energy between protons (i.e., spins) of opposite orientations is uniquely a T2 process, and although it is not the only cause of T2 relaxation, the term **spin-spin relaxation** is often used to refer to T2 relaxation.

In Chapter 9 we saw how the T1 relaxation time depends on the amount of magnetic "noise" at the Larmor frequency which in turn depends on the molecular motions (Fig. 9-1). The protons in molecules that are very mobile in fluids like water or CSF have a long T1, as do molecules that are relatively non-mobile such as those in membrane lipids and in macromolecules. Only molecules with intermediate mobility have short T1 relaxation times (Fig. 9-2).

The protons in molecules that are very mobile experience a

wide range of magnetic "noise" frequencies. This means that the amount of noise present at any specific frequency, such as the Larmor frequency or the lowest frequencies, is rather small. Both T1 and T2 will therefore be long (point c in Fig. 12-1). The protons in molecules with intermediate mobility experience magnetic "noise" that does not include the higher frequencies seen by more mobile molecules. This restriction in the frequency range results in more "noise" at the remaining frequencies, including the lowest frequencies and the Larmor frequency (assuming that the frequency range has not fallen below the Larmor frequency). This increased "noise" enhances the T1 and T2 relaxation processes and shortens both T1 and T2 (point b in Fig. 12-1).

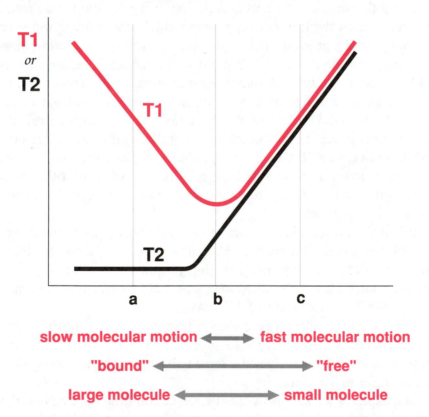

Fig. 12-1. The dependence of T1 and T2 on the frequency of molecular thermal motions. The points a, b, and c on this graph correspond to curves a, b, and c respectively in Fig. 9-1. With fast molecular motions both T1 and T2 are long. As molecular motions slow down both T1 and T2 at first decrease. With further slowing of molecular motions T2 continues to decrease, but T1 increases and becomes much greater than T2. Freely moving molecules have faster molecular motions than those that are bound; also, smaller molecules have faster molecular motions than larger molecules.

Protons in relatively non-mobile molecules will experience magnetic "noise" that is restricted to low frequencies and thus is very strong at these frequencies. The lack of "noise" at the Larmor frequency causes T1 to be very long, but the strength of the "noise" at low frequencies causes T2 to be very short (point a in Fig. 12-1).

The above explanations for the effect of molecular mobility on T2 relaxation can be restated in somewhat different terms. Highly mobile molecules see the local magnetic fields as quickly fluctuating. The effects of these local fields average out over a period of milliseconds, resulting in no significant differences in the average magnetic field seen by different protons. The local magnetic field fluctuations are therefore inefficient in producing proton dephasing and T2 relaxation. Protons in non-mobile molecules do not experience this averaging out of the local magnetic fields and thus different protons will experience different magnetic fields even when averaged over a period of milliseconds. The result is effective proton dephasing and a short T2 relaxation time.

Apart from cortical bone and possible calcifications, the human body is not usually thought of as containing solids. However, macromolecules incorporated into cell membranes and structures like myelin in nerve tissue can be considered solids in the sense that their molecules are relatively non-mobile. In these tissues the T1 relaxation time is long but the T2 relaxation time is very short; so short, in fact, that the signals from these molecules are lost even before the FID (or echo — see Chapter 13) can be measured. The signals (FID or echo) measured in MRI are therefore derived only from the **"mobile" protons**, i.e., protons belonging to relatively mobile molecules or components of molecules. This means that nearly all of the MR signal stems from the hydrogen nuclei in either water molecules or fat molecules (of the kind present in adipose tissue).

As discussed in Chapter 9, the mobility of water molecules and hence the relaxation properties of their protons is highly dependent upon the presence of other molecules. Macromolecules having hydrophilic side groups (e.g., proteins or parts of cell membranes) will attract water molecules and thus slow down their fast thermal motion, thereby decreasing both T1 and T2. Unbound lipids have intrinsically slower motions than water due to their larger size. Thus fat will also have decreased T1 and T2.

What effect does changing the strength of the static magnetic field, and thus changing the Larmor frequency, have on T1 and T2 relaxation times? The answer can be found in Figure 9-1. For protons in highly mobile molecules, changing the Larmor frequency will not significantly change the amount of magnetic "noise" present at the Larmor frequency, and thus neither T1 nor T2 will be much affected.

For protons in molecules with intermediate or low mobility, increasing the Larmor frequency will result in a decrease in magnetic "noise" at the Larmor frequency. This will cause a significant increase in T1 as the Larmor frequency is increased. What about the effect on T2? In most tissues we find that the T2 relaxation time is about 1/10 the T1 relaxation time at high field strengths and 1/5 of T1 at low field strengths. This means that the T2 relaxation rate is about 5 to 10 times the T1 relaxation rate. This would only be the case if the contribution to the T2 rate from low frequency "noise" was much greater than the contribution from the "noise" at the Larmor frequency (which determines the T1 rate). Since increasing the Larmor frequency has no effect on the amount of magnetic "noise" present at low frequencies, T2 will increase only slightly as the Larmor frequency in increased (for protons in molecules with intermediate or low mobility).

In explaining the effect of molecular mobility on magnetic "noise" we have assumed that most of this noise is produced by the neighboring protons and that the total amount of noise somehow summed over all frequencies is roughly constant. The addition of paramagnetic substances, however, changes these assumptions. The paramagnetic ion or molecule has a very large magnetic moment compared to the protons and greatly increases the strength of the local magnetic field fluctuations close to it. This increase in the strength of the magnetic "noise" at all frequencies will cause a decrease in both T1 and T2, though the effect on T1 will usually be more dramatic (Chap. 16).

13. The Spin-Echo Pulse Sequence, Separation of T2 from T2*

Having discussed the physical basis of T2 relaxation, we will discuss how to produce contrast in the MR images that depends on the differences in T2 between tissues. Previously we described a method that demonstrated the T1 differences in tissues. We now want a method in which the MR image brightness reflects the T2 relaxation in the tissues. In other words, the amplitude of the induced MR signal should be determined primarily by the rate of T2 relaxation, and should not be influenced by the dephasing of protons due to the inhomogeneities in the static magnetic field produced by the MR magnet. It is not possible to totally eliminate the static magnetic field inhomogeneities; what can be eliminated, however, is the effect of the static magnetic field inhomogeneities on the MR image. We can accomplish this with a technique that uses a **spin-echo pulse sequence**.

The spin-echo pulse sequence consists of a single 90° pulse followed by one or more 180° pulses. The effects of these pulses on the magnetization vector are shown in Fig. 13-1. This figure is depicted in the rotating frame of reference (see Ch. 6). As viewed from above (i.e., from a position on the positive z-axis), this frame of reference is rotating clockwise around the z-axis (the static magnetic field direction) at the "average" Larmor frequency $f_0 = (\gamma/2\pi)B_0$; B_0 is the average strength of the static magnetic field. The x and y axes of the rotating frame are labeled x' and y' to avoid confusion with the non-rotating axes.

The magnetic moments of protons precessing at this "average" Larmor frequency around the z-axis will appear stationary in the rotating frame. Only precessions having a different frequency or direction will appear to be precessing in the rotating frame.

In Figure 13-1a, a 90° pulse has just been applied. This RF pulse creates, for a short time, a magnetic field vector ($\mathbf{B_1}$) that rotates or precesses in the x-y plane at the Larmor frequency, f_0 (see Ch. 6). In the rotating frame of reference, this magnetic vector will be motionless. In Fig. 13-1a it has been placed along the x-axis. (It could have been placed in any direction in the x-y plane with the same effect.) Immediately after the 90° pulse, the magnetization that was originally along the z-axis is now in the x-y or transverse plane; specifically in this figure it lies along the y'-axis. At this time the transverse magnetization (M_{xy}) has its maximum size, because all the protons comprising the spin excess are pre-cessing around z together. This transverse magnetization produces the FID signal in the receiver coil. After a short time, however, the protons have started to dephase, i.e., their individual magnetic moments no longer point in the same direction, but have begun spreading out in the transverse plane (Fig. 13-1b). This causes the FID signal to decrease.

The three numbered arrows in the figure represent three such protons that have dephased. Proton 2 is precessing at exactly the same Larmor frequency as the rotating frame of reference (f_0); therefore the magnetic moment of this proton appears motionless in the figure and stays on the y'-axis. Proton 3 experiences a slightly higher magnetic field strength than B_0; therefore its preces-sion (in the clockwise direction) is slightly faster than f_0. The mag-netic moment of proton 3 has consequently rotated more degrees (in the clockwise direction) than that of proton 2. Proton 1, on the other hand, experiences a slightly lower magnetic field strength than B_0; its magnetic moment is precessing slightly slower than f_0 and is thus lagging behind the magnetic moment of proton 2. (Note: In the stationary frame of reference, proton 1 has not actu-ally precessed in an anti-clockwise direction; like all protons it is precessing clockwise, only more slowly than proton 2.)

Fig. 13-1 (facing page). Echo formation using a 180° pulse as viewed in the rotating frame. (a) The application of a 90° pulse rotates the magnetization into the x-y plane. (b) A short time after the 90° pulse the protons have totally dephased and the transverse magnetization, M_{xy}, has disappeared. The magnetic moment vectors of 3 selected protons are shown in the figure. These protons are precessing at slightly different frequencies. Proton 2 is precessing at exactly the same frequency as the rotating frame of reference, so that it appears stationary in this frame. Proton 3 is precessing slightly faster, and proton 1 is precessing slightly slower. (c) A 180° pulse is applied a time τ after the 90° pulse, creating the magnetic field vector, B_1, along the x' axis. This rotates each proton mag-netic moment by 180° around the x' axis. (d) The fastest precessing proton (3) is now behind the slowest precessing proton (1). (e) After another time interval τ (a total of 2τ after the 90° pulse) the faster precessing protons will have caught up with the slower precessing protons. At that time the proton magnetic moments will, for a moment, be precessing together (in phase) so that the trans-verse magnetization reappears. The longitudinal magnetization (M_z) is zero immediately after the 90° pulse. Due to longitudinal relaxation, M_z grows in the positive z direction. M_z just before the 180° pulse is shown in (b). The 180° pulse flips M_z by 180°; just after the 180° pulse M_z points in the negative z direction. This is shown in (d). (M_z is not shown in (c).) M_z is zero at (a) and practically zero at (e).

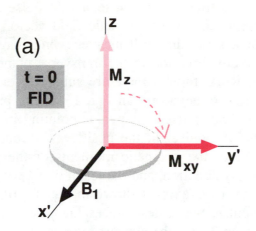

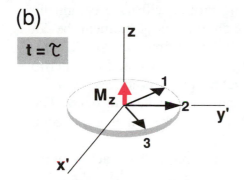

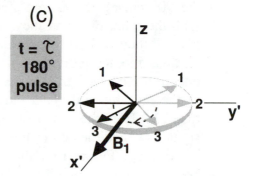

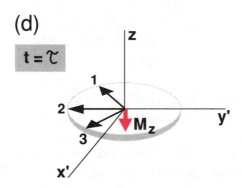

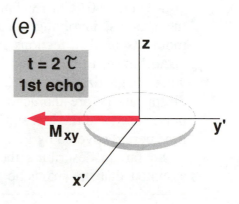

At a time τ after the 90° pulse, a 180° pulse is applied (Fig. 13-1c) so that a precessing magnetic field vector ($\mathbf{B_1}$) is again produced, for a short time.* This precessing magnetic field can be oriented in any direction in the transverse plane for the spin echo process to work, though there are subtle differences in net effect with different orientations. In Fig. 13-1c the precessing magnetic field of the 180° pulse is shown as pointing in the same direction as was the 90° pulse: in the positive x' direction. The magnetic field $\mathbf{B_1}$ forces the magnetic moments of the protons to precess around the $\mathbf{B_1}$ direction, and the result of a 180° rotation around $\mathbf{B_1}$ (i.e., around the x'-axis) is shown in Fig. 13-1d. Immediately after the 180° pulse, the fastest precessing proton (3) is now lagging behind proton 2, and the slowest precessing proton (1) is ahead of proton 2. However, at a time τ after the 180° pulse (i.e., 2τ after the 90° pulse), the fastest precessing proton will have caught up with the other two, and they will again, for a brief moment, be "in phase": they will all be pointing in the same direction (Fig. 13-1e). At this time, the individual magnetic moments will add together to once more create a transverse magnetization that induces a current in the receiver coil. This induced current is called an **echo**. The time (2τ) from the center of the 90° pulse to the center of the echo signal is called the **echo time (TE)**.

Although Figure 13-1 shows only three protons, the actual number of protons involved is, of course, enormous. For visual clarity, this figure does not show total dephasing of the three protons between the 90° and 180° pulses. In reality a total dephasing of all the protons, involving a complete fanning out of the magnetic moments in all directions in the transverse plane, does normally occur between the 90° and 180° pulses. This means that the transverse magnetization completely vanishes and that the FID signal disappears before the echo signal starts. Fig. 13-2 shows the RF pulses used and the MR signal detected in the receiver coil over the time period covered by Fig. 13-1.

So far, in describing the process of echo formation we have assumed that the magnetic field inhomogeneities that caused the

---

* We have referred to the magnetic field vector produced during both a 90° or 180° pulse as B_1. If the strengths of the oscillating magnetic field produced by the 90° and 180° pulse are in fact identical, then the 180° pulse will be twice as long in time as the 90° pulse. However, it is also possible for the 90° and 180° pulses to have the same time length, if the 180° pulse produces twice the magnetic field strength as the 90° pulse (see Equation 6-4).

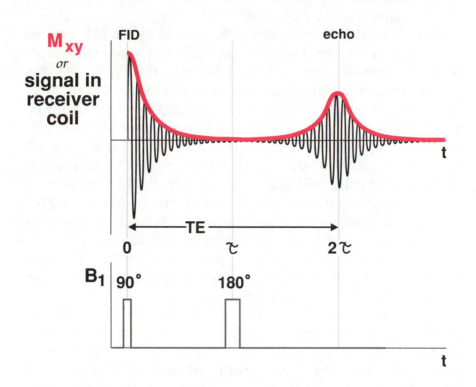

Fig. 13-2. Basic spin-echo pulse sequence using a 90° pulse and a single 180° pulse to produce a single echo. This figure shows the same sequence of events as Fig. 13-1. The application of a 90° pulse produces the FID which quickly disappears as the protons dephase. The application of a 180° pulse a time τ after the 90° pulse produces an echo a time 2τ after the 90° pulse. This time interval of 2τ is called the echo time, TE. The oscillating curve shown in black in this, and other, figures is the signal induced in the receiver coil. The amplitude, or envelope, of this signal is the same shape as the transverse magnetization, M_{xy}, which is shown in red.

protons to precess at different rates are constant in time. This means that the precessional frequencies of each of the protons in Fig. 13-1 are also constant in time. (Another assumption here is that the protons are stationary; there is no motion due to bulk flow or diffusion.) The result is a perfect alignment of the magnetic moments of the protons at the time of the echo. The size of the magnetization vector at the time of the echo is then the same as the size immediately after the 90° pulse. This means that the maximum amplitude of the echo signal is the same as the maximum amplitude of the FID signal. We have not yet, however, put in all the elements needed to complete our picture. In reality, the echo signal is always smaller than the FID signal.

The magnetic field inhomogeneities created by the MR magnet are in fact nearly constant over the echo times normally used. However, there are other sources of magnetic field variations that are not constant in time. These magnetic field variations or magnetic "noise" are produced by the tissue itself and are the source of the T2 relaxation process. These processes were discussed at length in Chapter 12. Since the T2 relaxation processes are intrinsically random, their effects on the protons' precession before and after the 180° pulse will not cancel; in fact, the 180° pulse will have no apparent effect on the dephasing that is due to T2 processes. As a result, at the time of the echo, the magnetic moments of the protons will not be perfectly in phase and the size of the transverse magnetization will be less than that immediately after the 90° pulse; the peak amplitude of the echo will be less than the peak amplitude of the FID, by a factor of $e^{-TE/T2}$.

In summary, *at the time of the echo*: (1) The size of the MR signal, ideally, is not reduced by the inhomogeneities of the static magnetic field. (2) The reduction in size of the MR signal, compared to the size of the FID, is determined solely by T2 relaxation. The 180° pulse does not in any way interfere with the signal reduction caused by T2 relaxation processes.

Instead of transmitting just a single 180° pulse, it is possible to transmit several 180° pulses, thereby obtaining several echoes of the original FID. Due to T2 relaxation, the maximum amplitude of successive echoes decays exponentially with a time constant equal to T2. In Figure 13-3 the line touching the "top" of the echoes is the T2 relaxation curve. Ideally, the amplitudes of these echoes will reflect the true T2 relaxation within the tissue, and will not be influenced by the field inhomogeneities caused by the MR magnet. In practice, the echoes can decay more rapidly than indicated by T2 relaxation if they are affected by the following phenomena: 1) bulk flow or diffusional motion of the protons (in addition to their thermal motions), or 2) imperfections in the application of the 180° pulses.

So far, by discussing the effect of the spin-echo pulse sequence on the transverse magnetization, we have only seen how the echo signal depends on T2 relaxation. In the next chapter we will see the effects of T1 relaxation on the echo signal. The contrast in an MR

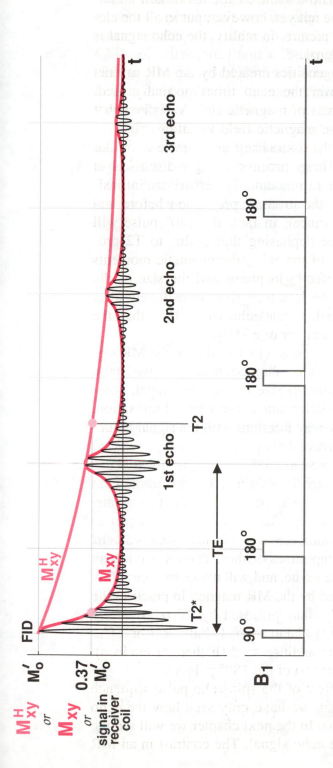

Fig. 13-3. Spin-echo pulse sequence using a 90° pulse and multiple 180° pulses to produce multiple echoes. This figure is similar to Fig. 13-2 except that several 180° pulses have been used to produce multiple echoes. Each 180° pulse will produce an echo at a time after the 180° pulse that is equal to the time between the previous echo or FID and the 180° pulse. In this figure, the time intervals between 180° pulses are all the same and are equal to twice the interval between the 90° and first 180° pulse. This particular spacing is not required; the spacing of each 180° pulse can be independently adjusted to any desired interval. In all cases a 180° pulse will exist at the midpoint of the interval between the preceding and following echo. In addition to the signal in the receiver coil, this figure also shows M_{xy} (dark red) and M_{xy}^H (light red). Since the effect of the magnetic field inhomogeneity produced by the MR magnet is eliminated at the peak of the echoes, a curve connecting the peak of the FID and the peaks of the echoes will show the relaxation that the transverse magnetization would have in a perfectly homogeneous magnetic field. This is, therefore, the T2 relaxation curve.

image produced using a spin-echo sequence depends on both transverse (T2) and longitudinal (T1) relaxation of the magnetization.

In the spin-echo pulse sequence, when the 180° pulse is transmitted at a time τ after the 90° pulse, a small magnetization (M_z) has been regained along the z-axis due to T1 relaxation (Fig. 13-1b). A rotation of this longitudinal magnetization by 180° around the vector $\mathbf{B_1}$, i.e., around the x'-axis, results in an equally small longitudinal magnetization in the antiparallel direction (Fig. 13-1d). After the 180° pulse, T1 relaxation begins again, and by the time the echo is induced, the magnetization along the z-axis has increased from a negative value to essentially zero (Fig. 13-1e).* With repeated 180° pulses, M_z is also essentially zero when subsequent echoes occur.

14. Image Contrast with the Spin-Echo Pulse Sequence: T2-Weighted, T1-Weighted, & Proton Density Weighted Images

Spin-echo is the pulse sequence of choice whenever image contrast dependent upon T2 is desired. This pulse sequence, however, is quite versatile; it can be adjusted to produce images in which the image contrast is sensitive to variations in T2, T1, or proton density. In this method, the FIDs are ignored and only the echoes are used for image reconstruction.

As mentioned in Chapter 7, one echo (or FID) does not contain sufficient information for image reconstruction. The production of even a single image requires a large number of successive spin-echo pulse sequences. Each spin-echo pulse sequence is composed of a 90° pulse, and one or more 180° pulses. An echo is produced following each 180° pulse (Fig. 14-1).

---

* Actually, M_z has a very small positive value at the time of the echo. Since T1 relaxation is exponential, the rate of change in the longitudinal magnetization is greater the farther it is from equilibrium. Thus the change in magnetization from a negative value after the 180° pulse is slightly greater than the change from zero that occurs after the 90° pulse.

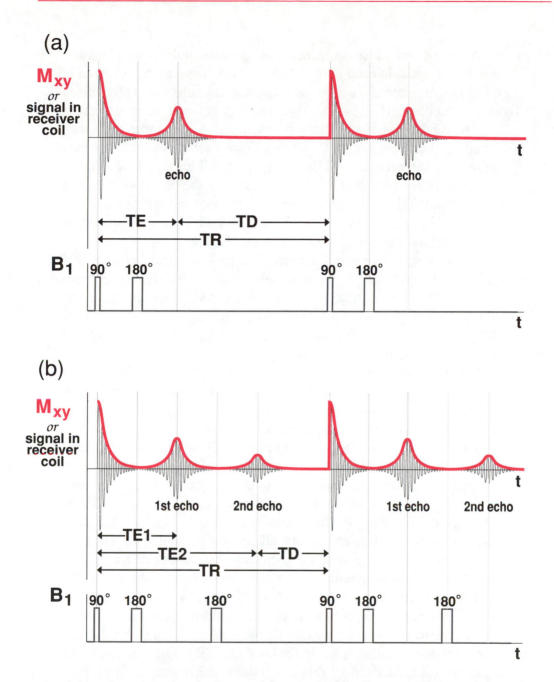

Fig. 14-1. Repeated spin-echo pulse sequences. The repetition time, TR, is the time between successive 90° pulses. (a) A 90°-180° pulse sequence producing a single echo. TE is the time between the 90° pulse and the peak of the echo. TD (delay time) is the time between the peak of the echo and the next 90° pulse. (b) A 90°-180°-180° pulse sequence producing two echoes. TE1 is the time between the 90° pulse and the peak of the first echo. TE2 is the time between the 90° pulse and the peak of the second echo. In this figure TE2 is shown as twice TE1, but this is not required; the selection of TE2 is totally independent of TE1. TD is the time between the last echo and the next 90° pulse.

A spin-echo pulse sequence can contain only a single 180° pulse and a single echo (Fig. 14-1a); in this case a single image is obtained. If, instead, each pulse sequence contains two 180° pulses and two echoes (Fig.14-1b), then two images can be constructed: one using all the first echoes in each pulse sequence, the other using all the second echoes. For this 2-echo sequence there are two values for the echo time (TE) which we call **TE1** and **TE2**. TE1 is the time from the center of the 90° pulse to the center of the first echo signal. TE2 is the time from the center of the 90° pulse to the center of the second echo signal. TE1 and TE2 are independently selectable; TE2 does not have to be twice TE1.

In the spin-echo pulse sequence, the time interval between 90° pulses is referred to as the **repetition time (TR)**. The time from the last echo of a pulse sequence to the next 90° pulse is called the **delay time (TD)**. For a single echo pulse sequence, TE + TD = TR. For a multiple echo pulse sequence, TEn + TD = TR, where TEn is the echo time of the last echo.

The versatility of spin-echo pulse sequences comes from control of the echo time (TE) and the repetition time (TR). By changing TE and TR, the operator of the MRI scanner can greatly affect the characteristics of the resultant image.

At the moment of the last echo in each pulse sequence, the longitudinal component of the magnetization, M_z, is nearly zero (Chap 13). After this echo, M_z in the various tissues increases according to the T1 relaxation rate in each particular tissue. If the delay time TD (and thus the repetition time TR) is long enough to allow complete T1 relaxation in all tissues present, i.e., TD is at least 4-5 times the longest T1, then any differences in the size of the longitudinal magnetization at the time of the next 90° pulse will reflect differences in proton density only. Since this longitudinal magnetization is then rotated into the transverse plane by the 90° pulse to become the source of the FID and echo signals, one can see that a sufficiently long TR (and TD) will minimize the effect of differences in T1 relaxation on image contrast.

If the pulse sequences also use a TE that is short compared to the T2 relaxation times of the tissues present, then little T2 relaxation will occur between the time of the 90° pulse and the echo (Fig. 14-2). In this case, differences in the echo amplitude from different tissues will be determined mainly by differences in proton density.

With a long TR (and TD) and a short TE, effects of both T1 and T2 relaxation on image contrast will be minimized; the image *contrast will be determined principally by the variations in proton density of the various tissues. Such an image is consequently called a **long TR, short TE** or **proton density weighted** image. In this type of image, the higher the proton density of a tissue, the brighter its appearance in the image.*

If, instead, the TE of the pulse sequence is adjusted so that it is longer — similar to the T2 relaxation times of the tissues — then significant T2 relaxation of the transverse magnetization will occur by the time of the echo (Fig. 14-3). The amplitude of the echo signal derived from a voxel of the tissue will then be reduced by the amount of T2 relaxation that has occurred there and thus will reflect both the proton density and the T2 relaxation in that voxel. Tissues with longer T2 and/or higher proton density will produce stronger echoes.

The "T2 contrast," i.e., the differences in echo amplitude from different voxels due to T2 relaxation, is highly dependent upon the TE used. The largest separation between the T2 relaxation curves of two different tissues occurs after a time close to the average of the T2 values in the two tissues (disregarding differences in proton density). If this time is chosen as the echo time, TE, then T2 contrast will be maximized (Fig. 14-4).

*By using a pulse sequence with a long TR (and TD) and a TE of sufficient length to allow good separation of the different T2 relaxation curves of tissues present, the image contrast will reflect differences in T2 and in proton density. Usually the effect of T2 on image contrast will be more dramatic than the contribution from proton density variations. Such an image is therefore called a **long TR, long TE** or **T2-weighted** image. In a T2-weighted image, tissues with a longer T2 will produce a stronger echo and will therefore appear brighter in the image. In addition, tissues with higher proton density will also appear brighter.*

Though the T2 contrast effects usually dominate the image, it is important to recognize that image contrast can sometimes vanish due to unfortunate combinations of contrast contributions from T2 and proton density. In some cases a tissue with a longer T2 and lower proton density may show the same image brightness as a tissue with shorter T2 and higher proton density. (There can also be interference in the image contrast from T1 effects, which we will discuss shortly.)

We can also produce images using a fairly short delay time (TD) in the spin-echo pulse sequence. As observed with the T2

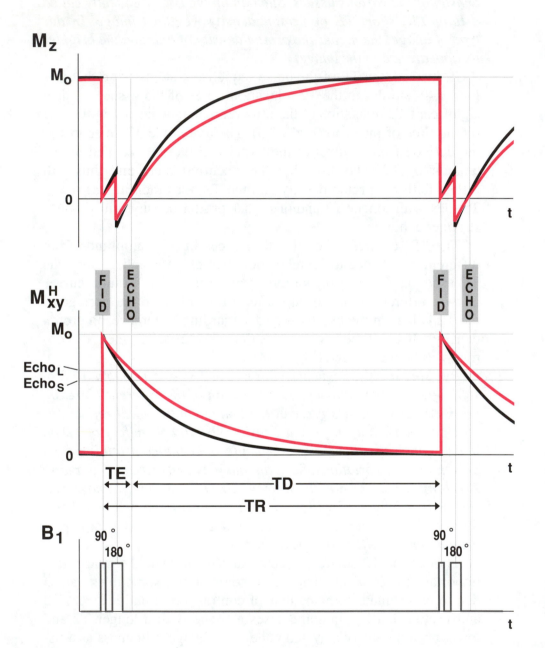

Fig. 14-2 (facing page). Proton density weighted, spin-echo pulse sequence. This figure (and Figs. 14-3 and 14-5) shows the variation in M_z and M_{xy}^H during a spin-echo pulse sequence that uses a 90° pulse and single 180° pulse. Two tissues with different relaxation times but identical proton densities are shown; the tissue with the longer relaxation times is represented by the red curves, the tissue with the shorter relaxation times is represented by the black curves. M_{xy} will equal M_{xy}^H at the times of peaks of the FIDs and the echoes; thus the size of M_{xy}^H at these times will give the relative sizes of the MR signal. $ECHO_L$ and $ECHO_S$ are the echo signals from the tissues with long and short relaxation times respectively.

This pulse sequence uses a long TR, so that TD is more than 4 times the T1 relaxation times of each of the tissues. It also uses a short TE; TE is as short as possible — shorter than the T2 relaxation times of the tissues. Since both tissues have experienced essentially complete T1 relaxation before the 90° pulse, the size of the MR signal will not depend on T1 relaxation. Since the TE is so short, ideally little T2 relaxation will have occurred at the time of the echo; thus T2 relaxation will have only a small effect on the size of the MR signal. These two tissues, with identical proton densities, yield almost identical echo signals. If these tissues had different proton densities, however, their echo signals would be proportional to their proton densities. Thus, this is called a proton density weighted pulse sequence.

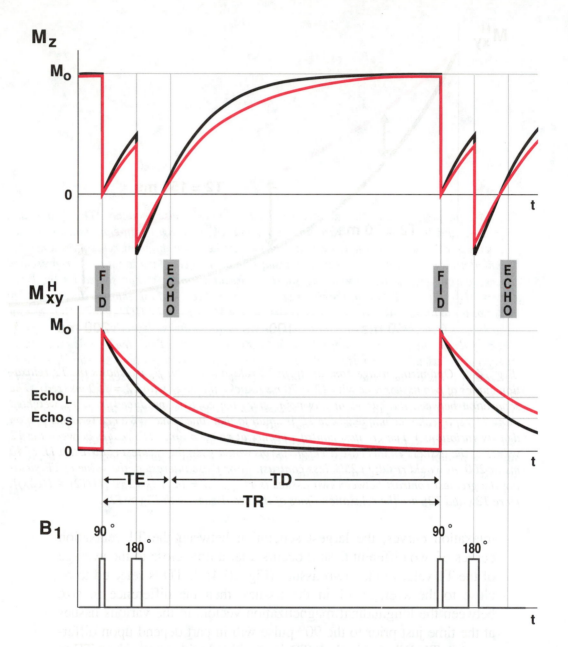

Fig. 14-3. T2-weighted, spin-echo pulse sequence. (Refer to the first paragraph of the caption for Fig. 14-2 for a description of the figure elements.) Like the pulse sequence shown in Fig. 14-2, the pulse sequence in this figure uses a long TR, so that TD is more than 4 times the T1 relaxation times of the tissues. However, TE is lengthened so that it is comparable to the T2 relaxation times of the tissues; substantial T2 relaxation will occur between the 90° pulse and the echo. Thus, the echo signals from the two tissue will reflect differences in T2 relaxation. Variations in proton density will also affect the echo signal in exactly the same way as in Figure 14-2, but the added sensitivity to T2 relaxation is usually more significant. Thus, this is called a T2-weighted pulse sequence.

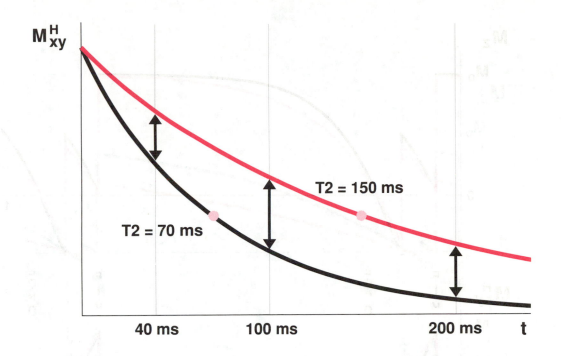

M^H_{xy}

T2 = 150 ms

T2 = 70 ms

40 ms   100 ms   200 ms   t

Fig. 14-4. Optimizing image contrast from T2 relaxation. This figure shows the T2 relaxation curves of two tissues, one with T2 = 70 ms (black), the other with T2 = 150 ms (red). The separation between the curves at any point gives the contrast produced by a T2 weighted (long TR) spin-echo technique whose TE is equal to that time value. (We are ignoring proton density variations.) The greatest contrast occurs at an echo time (TE) that is between the T2 values of the tissues. In this case a TE of 100 ms would yield the greatest contrast; a TE of 40 ms or 200 ms would result in 25% less contrast. (For those interested, the value of TE yielding the greatest contrast between two tissues is $TE_{optimum} = ln\ (T2_A/T2_B)/\ [(1/T2_B - 1/T2_A)]$. Here $T2_A$ and $T2_B$ are the relaxation times of the two tissues, with $T2_A > T2_B$.

relaxation curves, the largest separation between the T1 relaxation curves of two different tissues occurs after a time close to the average of the T1 values in the two tissues (Fig. 10-4). If TD is selected to be close to the average T1 in the tissues, then the difference in size between the longitudinal magnetization vectors in the various tissues at the time just prior to the 90° pulse will in part depend upon differences in T1. When this short TD is combined with a very short TE to minimize the effect of T2 variations, the differences in echo amplitude from the various tissues will be determined principally by differences in T1 and in proton density (Fig. 14-5).

In a pulse sequence that uses a very short TE and a TR (and TD) sufficiently short to allow good separation of the different T1 relaxation curves of tissues present, the image contrast will reflect differences in T1 and in proton density. Usually the effect of T1 on

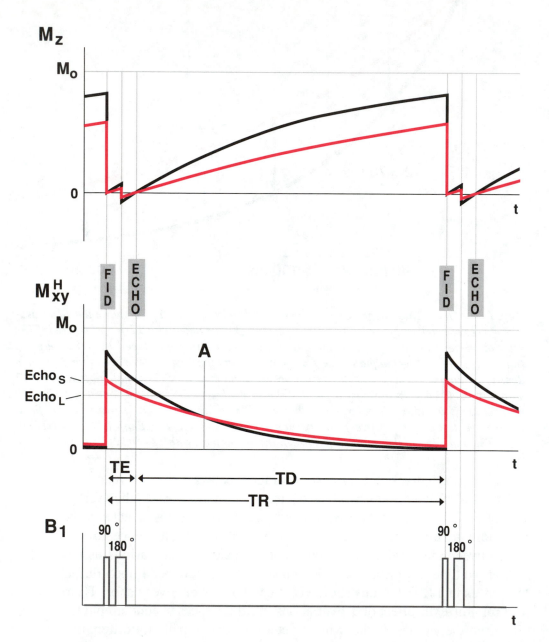

Fig. 14-5 (facing page). T1 weighted, spin-echo pulse sequence. (Refer to the first paragraph of the caption of Fig. 14-2 for a description of the figure elements.) Like the pulse sequence shown in Fig. 14-2, the pulse sequence in this figure uses a TE that is as short as possible to minimize the effects of T2 relaxation. However, TR is shortened so that TD is comparable to the T1 relaxation times of the tissues. At the time of the 90° pulse the T1 relaxation of the tissues will be quite incomplete, and the tissue with the longer T1 will have a smaller M_z; this will result in a smaller FID from this tissue and a smaller echo signal (provided TE is short). Thus the echo signals from the two tissues will reflect differences in T1 relaxation. Variations in proton density will also affect the echo signal in exactly the same way as in Fig. 14-2, but the added sensitivity to T1 relaxation is usually more significant. Thus, this is called a T1 weighted pulse sequence.

There is an unfortunate complication that can occur with this pulse sequence. This complication can eliminate the image contrast between tissues. Suppose TE is increased so that significant T2 relaxation occurs before the echo. M_{xy}^H in the fast relaxing tissue starts out greater than in the slow relaxing tissue. However, M_{xy}^H in the fast relaxing tissue will decrease more rapidly and eventually equal M_{xy}^H in the slow relaxing tissue. (This point is labeled "A" in the figure). If the echo occurs at this time of equal M_{xy}^H in the tissues, the MR signals from the tissues will be equal and image contrast will be lost. If the echo occurs after this time, the contrast of the tissues will be reversed with the faster relaxing tissue producing the smaller echo signal.

Pathological processes often cause an increase in T1 and T2 compared to normal tissue. If such a lesion (with lengthened relaxation times) is imaged using a relatively short TR, as above, its visibility and contrast in the image will depend upon the TE selected. For very short TE the lesion will appear darker (smaller echo signal) than normal tissue. As TE increases the lesion will disappear in the image. At still longer TE the lesion will again become visible, but with reversed contrast: it will appear brighter than normal tissue.

image contrast will be more dramatic than the contribution from proton density variations. Such an image is therefore called a **short TR, short TE** *or* **T1-weighted image.** *In a T1-weighted image, tissues with a shorter T1 will produce a stronger echo and will therefore appear brighter in the image. In addition, tissues with higher proton density will also appear brighter.*

Though the T1 contrast effects usually dominate in T1-weighted images, it is important to recognize that image contrast can sometimes vanish due to unfortunate combinations of contrast contributions from T1 and proton density. In some cases a tissue with a shorter T1 and lower proton density may show the same image brightness as a tissue with longer T1 and higher proton density. T2 relaxation can also have a significant unintended effect on image contrast. We will discuss this T2 effect shortly.

At this point you may wonder what the practical difference is between a T1-weighted spin-echo image and a saturation-recovery image that is also T1 sensitive. In the image reconstruction process there are certain advantages to having a complete signal that rises and falls in time like an echo, rather than a signal that starts at its maximum and declines as does the FID. In a sense, the FID is like the last half of an echo (actually a little less than half since the very peak of the FID is lost by interference from the end of the transmitted 90° pulse). Thus more complete data for imaging can be obtained from the echo than from the FID. There are also advantages in having a time interval between the 90° pulse and the MR signal; we shall discuss these in Chapter 19.

As indicated in Table 14-1, "T1-weighted," "T2-weighted," and "proton density weighted" are relative terms that may be true for some of the tissue components in the image, but usually not for all. For example, T1-weighted images utilize a short TE to minimize the amount of T2 relaxation that occurs before the echo. However, with any MRI scanner there is a practical limit beyond which TE cannot be reduced. In a T1-weighted image using a normal clinical setting for TE, some tissues with short T2s may still experience a significant amount of T2 relaxation during the TE interval. In like fashion, some tissues in a T2-weighted image may have T1 relaxation times that are greater than 1/4 of the selected TD. These tissues will not fully regain their maximum longitudinal magnetization during the TD interval.

In both the above cases, image contrast between tissues can be lost, since the image brightness is affected by both the T1 and T2

Table 14-1.

Image Type	Relative Length of:		Effects on Image Brightness of Tissues with:		
	TR	TE	Increased PD	Longer T1	Longer T2
Proton Density (PD) Weighted	Long[1]	Short[2]	+	(−)	(+)
T1-Weighted	Short[3]	Short[2]	+	−	(+)
T2-Weighted	Long [1]	Long[4]	+	(−)	+

In Table 14-1:

+: Indicates that the change in the indicated characteristic (proton density, T1 relaxation time, or T2 relaxation time) will cause the tissue to appear brighter in the image.

−: Indicates that the change in the indicated characteristic will cause the tissue to appear darker in the image.

. *(+) and (−): Indicate a possible, minor dependence of image brightness on these characteristics for some tissues.*

[1] *Ideally, TD is at least 4 times the T1 of the tissues present.*
[2] *TE as short as possible.*
[3] *TD selected to be approximately equal to the average of the T1 values of the tissues to be imaged.*
[4] *TE selected to be approximately equal to the average of the T2 values of the tissues to be imaged.*

characteristics of the tissues. This loss of image contrast is due to the following important facts:

(1) Tissues with a longer T1 usually also have a longer T2.
(2) For spin-echo imaging the effect of T1 and T2 changes on image contrast are opposing. Increasing T1 will make the image darker, while increasing T2 will make the image brighter.

Thus it is possible that a tissue with increased T1 and T2 may image with the same brightness as another tissue with shorter T1 and T2 if the image brightness for those tissues depends on both T1 and T2 (point A in Fig. 14-5).

Fig. 14-6 shows a proton density weighted image in the axial plane through the head of a patient with pressure hydrocephalus due to aquaductal stenosis. This MR image was obtained using a high field strength unit (1.5 T) with a long TR of 2,000 ms and short TE of 20 ms. This TR should assure that in the gray and white matter of the brain there is nearly complete reestablishment of the longitudinal magnetization vector during the TD interval. Thus the signal intensity between these two tissues should reflect the difference in proton density. The actual difference in proton density between these tissues is very small, however; much smaller than implied by the difference in their signal intensity. The low signal intensity of white matter is mainly due to its content of myelin. Due to very low molecular mobility, the protons in myelin have such a short T2 that they are not visualized (Ch. 12). The difference in signal intensity between gray and white matter is thus essentially due to a difference in the density of "mobile" protons (i.e., water).

The CSF seen in the lateral ventricles has a higher proton density than brain parenchyma, but still it is dark in this proton density weighted image. The T1 of CSF is, however, approximately 2 to 4 seconds (25, 29, 33, 58, 64). Thus, for CSF a TR of only 2 seconds

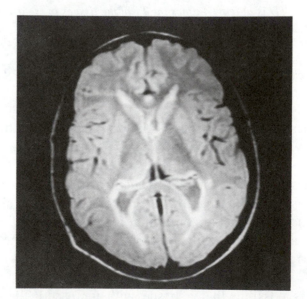

Fig. 14-6. Proton density weighted image (TR = 2,000 ms, TE = 20 ms) of a patient with pressure hydrocephalus. The bright rim surrounding the lateral ventricles is due to leakage of CSF, i.e., interstitial edema. Despite the term "proton density weighted" the contrast observed in the image is actually due to a combination of differences in proton density, T1 and T2.

would represent a T1-weighted (short TR, short TE) image, in which the CSF is dark due to its long T1. Some of the CSF-water is bright, however. Due to the pressure hydrocephalus, there has been leakage of CSF into the surrounding tissue. In this tissue some of the water molecules are slowed down and therefore the T1 of the water protons is shortened (Ch. 9). The periventricular edema consequently appears bright; brighter than CSF in the ventricles due to shorter T1 and brighter than brain parenchyma due to higher proton density. It should thus be apparent that the contrast in a proton density weighted image can be caused by differences in T1, T2 and proton density. The same will be true for T1- and T2-weighted images, but to a different extent.

The possibility of changing the contrast in an image by selectively emphasizing differences in proton density, T1 or T2, is very useful in clinical imaging. This can aid in the characterization of the various tissues and increase the detectability of pathological processes. Using only one set of pulse parameters (i.e., using only T1, T2 or proton density weighted images) increases the possibility of missing a lesion. Many pathological processes (neoplastic, inflammatory, ischemic and degenerative changes) have a long T1 and a long T2 as compared to the normal surrounding tissue (usually due to increased water content). It is quite possible to select the pulse parameters (i.e., the time intervals), such that the increased signal intensity provided by the long T2 is completely balanced by the reduced size of the magnetization vector due to the long T1, thereby providing no contrast whatsoever between the lesion and normal tissue. Multiple images derived from spin-echo sequences using multiple echoes can often provide needed variability in imaging parameters.

15. Inversion-Recovery: Maximum T1 Contrast or Additive T1 & T2 Contrast

Both saturation-recovery and spin-echo pulse sequences can produce images with T1 contrast. Maximum T1 contrast, however, is provided by a third type of pulse sequence called **inversion-recovery (IR)**. This pulse sequence is excellent for showing small differences in T1 between tissues.

The inversion-recovery pulse sequence begins with a 180° pulse (also called an **inversion pulse**) that rotates (or inverts) the magnetization vector. For the time being, let us assume that the magnetization is at equilibrium, with a size of M_0, just before the 180° pulse is applied. The 180° pulse then rotates the magnetization from its equilibrium orientation along the positive z-axis to an orientation along the negative z-axis, i.e., in the antiparallel direction (Fig. 15-1a). The protons are then allowed to relax. Since there is no magnetization vector in the transverse plane, no FID signal is produced and only T1 relaxation will take place. As the surplus number of antiparallel protons is gradually transformed into a surplus number of parallel protons, the magnetization vector along the z-axis increases from its maximum negative value, $-M_0$, through zero (when the number of antiparallel and parallel protons are equal) towards its maximum positive value at thermal equilibrium, $+M_0$ (Fig. 15-1b, c, d).

The change of the magnetization vector toward its equilibrium value is exponential, i.e, with an ever decreasing rate (Fig. 15-2). The part of the T1 relaxation curve above the zero line (i.e., for values of $M_z > 0$) is identical to the T1 relaxation curve after a 90° pulse (Fig. 8-2). The full range of the magnetization vector is $2M_0$ (from $-M_0$ to $+M_0$). T1 corresponds to the time it takes for the magnetization to undergo 63% of its total change, from $-M_0$ to $+M_0$. Thus, after a time T1, the magnetization vector will point in the positive z direction and will be equal to $0.26\ M_0$ (see below). After a time equal to 4-5 times T1, the magnetization vector has, for all practical purposes, completely regained its equilibrium size and orientation. It is also of interest to know that the length of time required for the magnetization to reach zero amplitude is 0.69 T1.

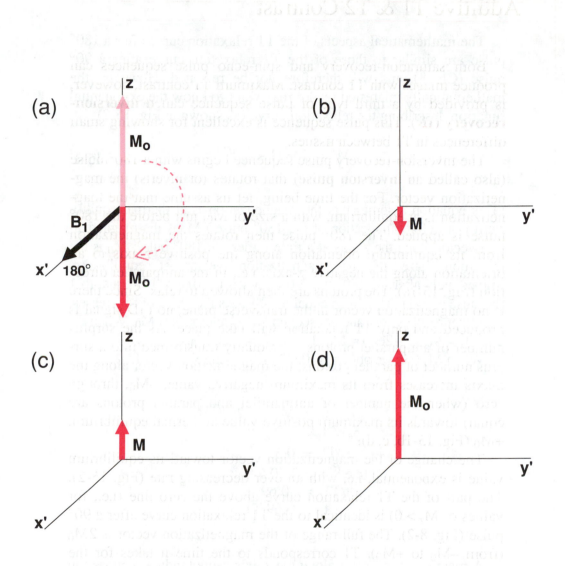

Fig. 15-1. Recovery of the equilibrium magnetization after inversion using a single 180° pulse (viewed in the rotating frame). (a) In this figure the magnetization is initially at thermal equilibrium. The application of a 180° pulse rotates (or inverts) the magnetization vector from the positive z to the negative z direction without changing its size. (b) and (c) As time elapses, the magnetization changes towards its original state. It first decreases in size until it reaches zero and then increases in the positive z direction. (d) Eventually, the magnetization reaches its original state of thermal equilibrium. At no time does a transverse magnetization exist, so a MR signal is never produced.

Mathematical Interlude

The mathematical aspects of the T1 relaxation curve after a 180° pulse are similar to those of the T1 relaxation curve after a 90° pulse (Ch. 8), and will therefore not be repeated. However, the mathematically interested reader may want to know the exponential function describing the relaxation curve (as shown in Fig. 15-2):

$$M_z = M_0 (1 - 2e^{-t/T1})$$ (15-1)

(This equation assumes that the magnetization is at thermal equilibrium just before application of the 180° pulse.) M_z is the magnetization vector along the z-axis at a time t after the 180° pulse; M_0 is the maximum size of the magnetization vector at equilibrium. Immediately after the 180° pulse (i.e., when t = 0) $e^{-t/T1}$ is equal to 1 (because $e^0 = 1$). M_z is then equal to $-M_0$. As the value of t increases, $e^{-t/T1}$ approaches zero and consequently M_z approaches M_0. When t = T1, $M_z = M_0 (1 - 2e^{-1}) = M_0 (1 - 2/2.7) = 0.26 M_0$.

When $M_z = 0$, we obtain t = 0.69 T1 by using Equation 15-1:

$$0 = M_0 (1 - 2e^{-t/T1})$$
$$2e^{-t/T1} = 1$$
$$e^{-t/T1} = 0.5$$
$$\ln e^{-t/T1} = \ln 0.5$$
$$-t/T1 = -0.69$$
$$t = 0.69 T1$$

A magnetization vector along the z-axis cannot induce a current in the receiver coil. We need to rotate this magnetization vector into the transverse plane in order to detect a signal. In the inversion-recovery pulse sequence, a 90° pulse is added at a certain time to accomplish this rotation of the magnetization vector from a longitudinal to a transverse orientation (Fig. 15-3). The time interval between the 180° pulse and the 90° pulse is called the **inversion time (TI)** (Fig. 15-4).

The amplitude of the FID that follows the 90° pulse will depend upon the size of the magnetization vector that is rotated into the x-y plane and this depends, in turn, on TI. If the inversion time is either

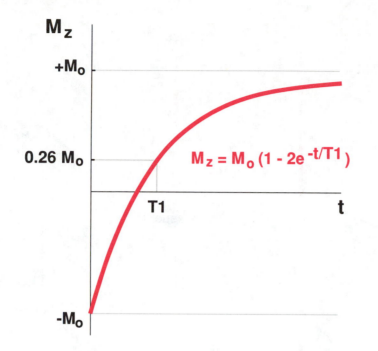

Fig. 15-2. T1 relaxation after inversion of the magnetization. Suppose that the tissue is originally at thermal equilibrium. After the application of a 180° pulse, the longitudinal magnetization (M_z) changes exponentially from $-M_0$ to $+M_0$; t is the time after the 180° pulse. This relaxation curve is described by the following exponential function: $M_z = M_0 (1 - 2e^{-t/T1})$. At a time T1 after the 180° pulse, M_z will have relaxed 63% of the way from $-M_0$ to $+M_0$; thus, M_z will equal $+0.26 M_0$ (and point in the positive z direction).

zero or at least 4-5 times the T1 of the tissue, maximum signal amplitude will be obtained, since the magnetization will be at its maximum, M_0, during the 90° pulse. If the inversion time is 0.69 T1, no FID signal will be produced since the magnetization will be zero when the 90° pulse is applied.

With inversion-recovery it is very important to realize that maximal image contrast does not occur with the inversion times that produce maximal FID signals. In fact, with the very short or very long inversion times that produce the largest FID signals, image contrast is at a minimum. However, if the inversion time is set at approximately the average T1 value of two tissues, maximum image contrast between the two tissues will result.

This value for TI that produces maximum contrast in the inversion-recovery method is identical to the value for TR in the saturation-recovery method (see Fig. 10-4). However, due to the fact that the

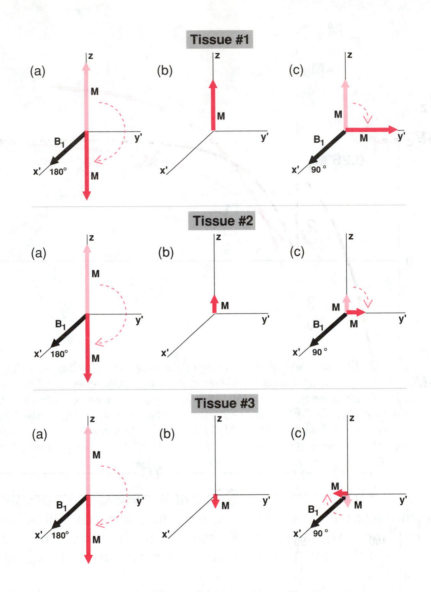

Fig. 15-3. The inversion-recovery pulse sequence as viewed in the rotating frame for 3 tissues with different T1s. In this figure (and in Fig. 15-4) tissues with 3 different T1 relaxation times are shown. Tissue #1 (shown in the top row of figures) has the shortest T1, while tissue #3 (shown in the bottom row of figures) has the longest T1. The T1 of tissue #2 is intermediate, between that of #1 and #3. (a) The application of a 180° pulse rotates the magnetization from the positive z direction to the negative z direction. (b) Just before the application of a 90° pulse the magnetization in all tissues has relaxed somewhat toward the equilibrium value. The magnetization has changed the most in tissue #1 and the least in tissue #3; in fact, the magnetization in these two tissues now points in opposite directions. (c) The application of a 90° pulse flips the magnetization into the x-y plane, so that a FID is produced. Note that the transverse magnetization of tissue #1 is now pointed in the opposite direction from that of tissue #3 (they are 180° out of phase).

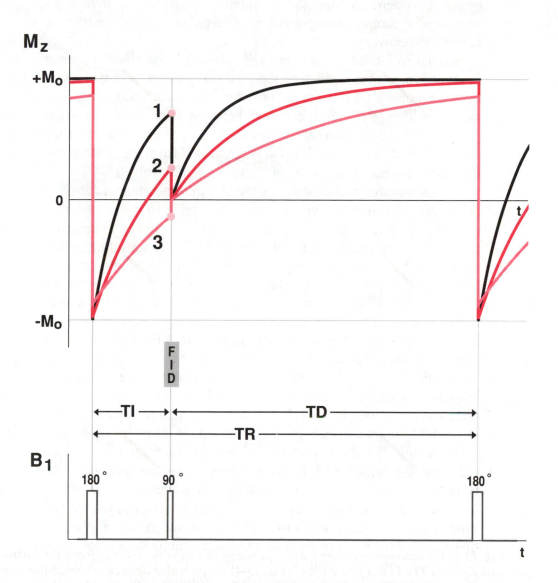

Fig. 15-4. The inversion-recovery pulse sequence. This figure shows a plot of the longitudinal magnetization for the same 3 tissues as Fig. 15-3. M_0 is the equilibrium magnetization. TI (the inversion time) is the time between the 180° inversion pulse and the 90° pulse. TD (the delay time) is the time between the 90° pulse and the next 180° inversion pulse. TR (the repetition time) is the time between successive 180° inversion pulses. The size of the FID signal that occurs immediately after the 90° pulse is given by the size of M_z immediately before the 90° pulse (this is indicated by the red dots).

longitudinal magnetization vector in inversion-recovery has potentially twice the range of that in saturation-recovery ($-M_0$ to M_0 as opposed to zero to M_0), the potential T1 contrast in inversion-recovery is correspondingly twice as large as the T1 contrast in saturation-recovery.

Similar to saturation-recovery and spin-echo, the pulse sequence in inversion-recovery has to be repeated many times to obtain a sufficient number of FIDs (or echoes) for image reconstruction. The repetition time (TR) is the period between each 180° inversion pulse, and this time interval is equal to the sum of the inversion time (TI) and the delay time (TD). TD is the period from the peak of the FID to the middle of the next 180° inversion pulse (Fig. 15-4).

So far we have assumed in this chapter that the magnetization is at thermal equilibrium, with a size of M_0, prior to the application of the 180° pulse. In actual practice this may only be approximately true. During the time delay interval (TD) longitudinal relaxation continues. Immediately after the 90° pulse the longitudinal magnetization starts with a value of zero and increases toward the value M_0. Essentially complete longitudinal relaxation occurs after a time equal to 4 to 5 times T1, at which time the longitudinal magnetization is nearly M_0. If TD is less than 4 to 5 times T1, the longitudinal magnetization will be less than M_0 just prior to the application of the 180° pulse. This will reduce the size of the subsequent MR signal. Since tissues with different T1s will experience different reductions in the MR signal, it may also interfere with the desired T1 contrast effects produced during the TI interval.

Reducing TD does have the advantage of reducing the time required to produce an MR image using this pulse sequence. To achieve this reduction in imaging time a slight reduction of signal intensity can be tolerated. In practice a time delay of at least 2 to 3 times the T1 of the tissues of interest in the image will allow adequate reestablishment of the longitudinal magnetization prior to the 180° inversion pulse and will not significantly interfere with T1 image contrast. If such a time delay is followed by an inversion time close to the average T1 of the tissues of interest, an MR image optimized for T1 contrast will be produced.

An interesting characteristic of inversion-recovery is the contrast possibilities provided by an inversion time short enough so that the longitudinal magnetization vectors in some tissues are still negative when the 90° pulse is transmitted (Fig. 15-3, tissue #3; Fig. 15-4, curve 3). When these negative M_z vectors are rotated

90° into the x-y plane, their orientation in that plane will be opposite that of positive M_z vectors rotated 90° into the x-y plane. In other words, a positive M_z vector and a negative M_z vector will have a phase difference of 180° in the x-y plane after a 90° pulse. If these vectors have the same absolute size, their induced FIDs will have the same amplitude, but differ in phase by 180°.

The image brightness assigned to these signals depends upon the image reconstruction method used (Fig. 15-5). There are two possibilities. A **phase corrected image reconstruction** requires a phase-sensitive signal detector; it assigns a dark shade of gray to the signal caused by the negative M_z and a bright shade of gray to the signal caused by the equally large, but positive M_z. No FID signal at all — corresponding to an intermediate T1 relaxation rate — is given an intermediate shade of gray. In other words, phase corrected reconstruction assures an accurate portrayal of T1 contrast for the full range of M_z values, both positive and negative.

The alternative image reconstruction method, called **magnitude image reconstruction**, does not distinguish between signals of opposite phase. The brightness in the image is determined by the signal amplitude only. If the magnetization vector in a tissue is zero when the 90° pulse is transmitted, no FID is induced and the tissue appears dark. Two tissues, having magnetizations that are equal in size but opposite in sign at the time of the 90° pulse, will image with the same shade of grey; they will not be distinguishable. These same tissues *would* be distinguishable using the phase correction method. It follows that a magnitude reconstructed inversion-recovery image made with a relatively short inversion time can have quite a confusing appearance.

The inversion-recovery pulse sequence with a very short inversion time (TI << T1) is of special interest because, when combined with a 180° echo pulse, it offers a feature not provided by any other pulse sequences: additive T1 and T2 contrast (Fig. 15-6). (Remember from Chapter 14 that whenever both T1 and T2 relaxation contributed to the image contrast in a spin-echo image, the effects of T1 and T2 on image brightness were opposing.) When a fairly long delay time (TD) is followed by a very short inversion time (TI), the longitudinal magnetization in all tissues will still be negative when the 90° pulse is transmitted. The tissue having the longest T1 will have the largest magnetization vector pointing in the negative z direction at the time of the 90° pulse. The largest FID signal will be produced by the tissue with the longest T1, the smallest FID by the tissue with the shortest T1. At the echo time (TE) the difference in the sizes of the echoes between the tissues with long and short T1 will be enhanced over the FID difference, provided that the tissue with the longest T1 also has the longest T2 (which is usually the case). Both the long T1 and the long T2 will thus contribute to give this tissue a larger echo amplitude than tissues having shorter T1 and T2. This unique inversion-recovery, spin-echo pulse sequence has been dubbed **STIR (short TI inversion-recovery)** (11, 64). Problems concerning possible loss of image contrast due to the opposing image brightness contributions from T1 and T2 that can occur in spin-echo images are much less likely in the STIR technique.

Actually, the addition of a 180° echo pulse to the inversion-recovery pulse sequence is the rule rather than the exception; this applies to all variants of inversion-recovery. The main reason is a practical one. It is difficult to measure a relatively weak FID immediately after the transmission of a strong 90° pulse, especially if the same coil is used for both transmission of the pulse and reception of the signal. This technical difficulty is circumvented by the introduction of a 180° echo pulse. The FID is ignored, and only the echo is detected.

This provides a small time interval between the 90° pulse and the registration of the signal (the echo), an advantage that will be discussed in Chapter 19. Using the echo rather than the FID signal has an additional advantage. The echo contains both a rising and falling side while the FID contains only a falling side; in a sense, the FID corresponds to the later half of an echo. This extra information contained in the echo is useful in the image reconstruction process.

When the 180° echo pulse, for practical purposes, is added to an inversion-recovery pulse sequence designed to give maximum T1 contrast, i.e., with an inversion time close to the average T1, the echo time should be kept as short as possible. Any added T2 contrast would oppose the T1 contrast and the result could easily become as shown in Fig. 15-7.

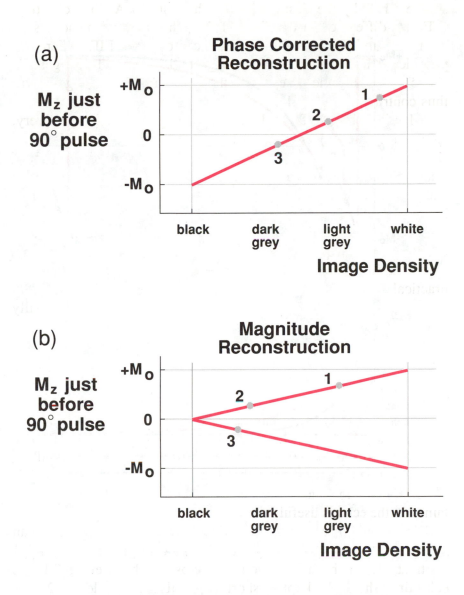

Fig. 15-5. "Phase corrected" versus "magnitude" image reconstruction. These graphs show the image density or brightness that results from different values of M_z immediately before the 90° pulse. (a) Phase corrected reconstruction can distinguish all values of M_z from $-M_0$ to $+M_0$. (b) Magnitude reconstruction is only sensitive to the magnitude of the magnetization, not to its direction. Thus two tissues with different T1s and with longitudinal magnetizations that are equal in size but opposite in direction, immediately before the 90° pulse, will image with the same shade of grey. Results from tissues #1, #2, and #3 of Fig. 15-3 and 15-4 are plotted in both (a) and (b). By using phase corrected reconstruction all 3 tissues should be easily distinguishable; by using magnitude reconstruction tissues #2 and #3 give nearly the same image density and will be difficult to distinguish.

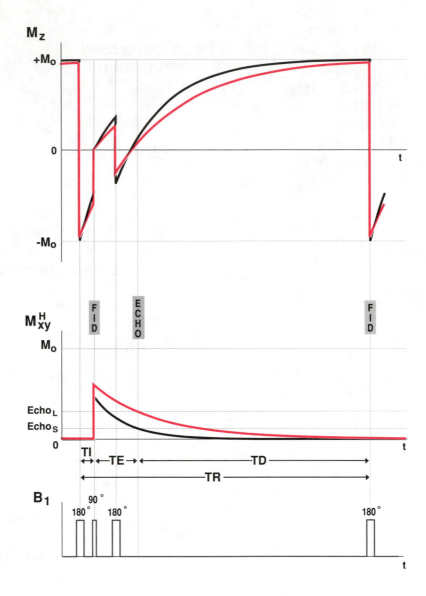

Fig. 15-6. Inversion-recovery, spin-echo pulse sequence with short inversion time showing additive T1 and T2 contrast: STIR (short T1 inversion-recovery). Two tissues are shown here, one with a shorter T1 and T2 (black), the other with a longer T1 and T2 (red). After a 180° inversion pulse, the 90° pulse is applied when M_z is still negative. In this case, the fastest relaxing tissue (shortest T1) will have the smallest FID. A 180° echo pulse is applied, producing an echo signal at a time TE after the 90° pulse. At the time of the echo, the echo signals from all tissues will be smaller than their FID signals. However, the tissue with the shortest T2 will experience a greater decrease in signal than the tissue with the longest T2. Thus the separation in signal amplitude between the fast relaxing and slow relaxing tissue will be enhanced in the echo compared to the FID.

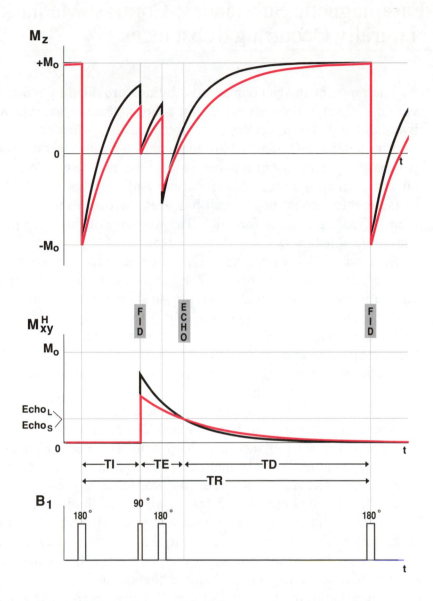

Fig. 15-7. Loss of image contrast using an inversion-recovery, spin-echo pulse sequence with longer inversion time. This figure is similar to Fig. 15-6 except that TI is longer. Here the longitudinal magnetization in some tissues has relaxed from a negative value to a positive value; two such tissues are described in this figure. In this case T1 and T2 contrast effects are no longer additive, but can cancel out, leaving no image contrast between certain tissues (even if phase corrected reconstruction is used). In this figure the tissue with the shorter T1 and T2 is black, while the tissue with the longer T1 and T2 is red.

In the FID the tissue with the shorter T1 will have the larger signal. However, this tissue will usually also have a faster T2 decay so that at the time of the echo it may produce the same size echo signal as another tissue with quite larger T1 and T2 values.

16. Paramagnetic Substances: Contrast Media & Naturally Occurring Substances

The range of image contrast possibilities provided by the many variants of saturation-recovery, spin-echo, and inversion-recovery is undoubtedly large, yet there may be occasions when differences in proton density, T1 and T2 are too small to allow separation of a pathologic process from the surrounding normal tissue. With the aim of increasing diagnostic sensitivity and specificity of MRI, a considerable amount of research has been devoted to the development of contrast media for MRI. The majority of these are **paramagnetic substances**.

So far in this book we have concentrated our attention on the magnetism produced by nuclei. The specific nucleus we have discussed is the hydrogen nucleus or proton — the one normally used in MRI. This "nuclear magnetism" is due to the magnetic moments of the components of the nucleus, the protons and neutrons (or in the important case of hydrogen, the single proton). To understand the actions of paramagnetic materials we must first look at the nature of atomic magnetism.

The electrons which surround the nucleus, like the proton, are also spinning charged particles, though the electron is negatively charged and is much lighter than the proton. The electron's mass is almost 2000 times less than that of a proton. Since the electron has both spin and charge it has, like the proton, a magnetic moment and will precess when placed into a magnetic field. In fact, the electron's intrinsic magnetic moment is approximately 700 times larger than the proton's, mainly due to the electron's much smaller mass. When placed into the same magnetic field the electron will precess at a frequency about 700 times that of a proton.

In addition to the magnetic moment of the electron itself, there is an additional contribution to the total magnetic moment of the atom due to the orbital motion of the electron around the nucleus. It is clear that the net magnetic moment of the atom can be dominated by the contribution from the electron. The magnetism produced by the spin and orbital motions of the electron is referred to as **atomic magnetism**.

In actuality, most stable ions, atoms, and molecules do not exhibit the strong magnetic properties that would be expected from the strength of the electron's magnetic moment. In many cases, the spins and orbital motions of the electrons in these substances pair off so that their magnetic moments cancel. We know that, to obtain stability, atoms can gain, lose, or share valence or outer electrons to form ions or covalent molecules. In stable atoms (noble gases), and stable ions (e.g., Na+, K+, Ca++, F–, Cl–) the outer electron sub-shells are complete. Both the spins and the orbital motions of the electrons in a completed subshell will normally pair off to yield a net magnetic moment of zero for the electrons in the subshell. In stable covalent molecules and metallic crystals the sharing of elec-trons also generally results in a cancelling of the spin and orbital magnetic moments of the outer electrons. (A notable exception is molecular oxygen, O_2, which does retain a net magnetic moment.)

There are elements, however, that contain *inner* electron sub-shells which are not complete (or full). Since these are inner, and not valance, subshells they do not participate in ionic and covalent bonding. They remain unfilled when stable chemical compounds are formed from these atoms. Elements with unfilled inner sub-shells include transition metals such as chromium, manganese, iron, cobalt, and copper, and lanthanide metals such as gadolinium, dysprosium, and europium. The unpaired electrons in these ele-ments will result in a net magnetic moment in the atoms, or in the ions or molecules that they form. We will refer to the magnetism produced by these atoms, ions, or molecules generically as **atomic magnetism**. We will also refer to the microscopic magnetic dipoles produced by the electrons as atomic magnetic dipoles, even though we may sometimes be referring to ions or molecules.

In many substances with atomic magnetic dipoles, the magnetic interactions between the atoms are not very strong and the individ-ual atomic moments act relatively independently. In the absence of a magnetic field the orientations of the atomic magnetic moments will be random; the bulk substance will show no magnetism. If the substance is placed in a magnetic field, however, the atomic mag-netic moments will tend to align with the magnetic field and pro-duce a magnetization vector in the direction of the magnetic field.*

---

* In this way atomic magnetic moments behave much like nuclear magnetic moments. The amount of alignment with the magnetic field direction is described by the same Boltzmann equation as used for protons (Eq. 3-1). However, since the strength of atomic dipoles is about 1000 times greater than nuclear dipoles, the ΔE in the Boltzmann equation is about 1000 times larger for atomic dipoles; their alignment with the magnetic field is enhanced by a factor of about 1000 over nuclear dipoles. Atomic dipoles experience thermal interac-tions with the tissue so that the orientation of individual dipoles constantly changes, just like nuclear dipoles.

This magnetization will produce in the material an additional magnetic field that is in the same direction as the applied field. Such a substance is termed **paramagnetic**. The ratio of the induced magnetic field produced by the substance to that of the applied magnetic field is a constant for the material and is called its **magnetic susceptibility**. The susceptibility is positive for paramagnetic materials; this simply means that the induced and applied magnetic fields point in the same direction.

In some materials with atomic magnetic moments, the magnetic interaction between atoms is strong enough that their magnetic moments act cooperatively. If a magnetic field is applied to such a substance the magnetic moments can be aligned; the degree of alignment will be greater than with paramagnetic materials. If the applied magnetic field is reduced to zero, many of the magnetic moments will remain aligned due to their cooperative interactions. The substance will now have a permanent magnetization and produce a magnetic field even in the absence of an applied external field. Such a substance is called **ferromagnetic**. Elemental nickel and iron are examples of such materials.

Superparamagnetic materials, like ferromagnetic materials, experience strong alignment of their magnetic moments by an external magnetic field; thus their magnetic susceptibility is high. However, they do not retain their magnetization when the external magnetic field is removed; in this way they are more like paramagnetic materials.

Substances that do not contain atomic magnetic dipoles (in the absence of an applied magnetic field) will still demonstrate a weak form of magnetism called **diamagnetism**. When such a material is placed into a magnetic field, the orbits of its electrons will be slightly affected; the result is a small induced magnetic moment in each atom that produces a magnetic field *opposed* to the applied field. The atomic magnetic moments induced by diamagnetism are generally thousands of times less than the permanent magnetic moments possessed by paramagnetic atoms, ions, and molecules. The magnetic susceptibilities of diamagnetic materials are negative.

Strictly speaking, the nuclear magnetization induced when protons are placed in a static field is also a type of paramagnetism — in this case termed **nuclear paramagnetism**. The atomic magnetization that is produced by unpaired electrons could be called **electronic paramagnetism**, but the word "electronic" is usually omitted.

Paramagnetic substances can affect the signal intensity and contrast in MR images by decreasing both the T1 and T2 relaxation times of the tissues in which they reside. The paramagnetic substance itself is not "visible": the atomic magnetic dipoles or **paramagnetic centers** do not directly contribute to the MR signal since their Larmor frequency is much different than the Larmor frequency of protons. However, their strong local magnetic fields affect nearby protons, substantially increasing their relaxation rates; this effect can be quite apparent in the MRI image.

In Chapters 9 and 12 we saw that T1 and T2 relaxation are due to the presence of magnetic "noise" — fluctuations of the local magnetic fields in the tissues. T1 relaxation is produced by "noise" at the Larmor frequency, while T2 relaxation is caused by both the low frequency component of the "noise" and the component at the Larmor frequency. Magnetic "noise" is produced by the local magnetic fields from nuclear and atomic magnetic dipoles in the tissue. The frequency components of the actual "noise" experienced by a proton are determined by the motions of the proton and by the motions and strengths of the magnetic dipoles near it. Another significant contribution to this noise, which comes from the paramagnetic centers, is the rapidly fluctuating magnetic field due to the flipping of the atomic dipole between the parallel and antiparallel orientations. These flips occur extremely rapidly compared to those of the protons. (The longitudinal relaxation times for the atomic dipoles are much less than 1/1,000,000 of the proton's typical relaxation time.)

The strength of the atomic dipoles is greater than that of the proton dipoles by about a factor of 1000. This strength is the reason that paramagnetic centers can be very effective at inducing relaxation of the protons. The magnetic "noise" and therefore the relaxation effects produced by a magnetic dipole, however, are mostly felt over a very short range — just a few angstroms. For effective relaxation, the proton must be able to approach closely to the dipole. The paramagnetic center itself is often a metal ion coupled to a large carrier molecule (ligand). The requirement of accessibility of water molecules to the paramagnetic ion thus implies some restrictions in the structure of the ligand.

In Chapter 11 we saw that when relaxation is produced by separate mechanisms we can obtain the net relaxation *rate* by adding the relaxation *rates* due to each of the separate mechanisms. When paramagnetic substances are added to a tissue, the resultant relaxation rate is the original relaxation rate, without the paramagnetic substance, plus a relaxation rate produced by the interactions of the protons with the paramagnetic centers:

$$\frac{1}{T1} = \frac{1}{T1'} + \frac{1}{T1_p} \qquad (16\text{-}1)$$

$$\frac{1}{T2} = \frac{1}{T2'} + \frac{1}{T2_p} \qquad (16\text{-}2)$$

$1/T1$ and $1/T2$ are the observed relaxation rates in the presence of the paramagnetic material; $1/T1'$ and $1/T2'$ are the original relaxation rates of the tissue without the paramagnetic material; $1/T1_p$ and $1/T2_p$ are the additional relaxation rates due solely to the paramagnetic centers.

$1/T1_p$ and $1/T2_p$ are not equal, but if they are roughly similar, it is easy to show that the effect on the T1 relaxation time is much more dramatic than the effect on T2.

In tissues, T1 is usually much longer than T2. Suppose that in a particular tissue T1 is 500 ms and T2 is 50 ms. The corresponding relaxation rates ($1/T1$ and $1/T2$) are 2 s^{-1} and 20 s^{-1}, respectively. An increase in both relaxation rates of 1 s^{-1} (caused by the paramagnetic substance), increases the T1 relaxation rate to 3 s^{-1} and the T2 relaxation rate to 21 s^{-1}. These rates correspond to a T1 of 333 ms and a T2 of 48 ms. T1 has decreased by 33%, while T2 has decreased by only 4%.

In general, a given paramagnetic concentration will produce a larger relative shortening of longer relaxation times. Therefore, among tissues having equal concentrations of paramagnetic material those with longer relaxation times will demonstrate the greatest changes in image brightness, i.e., the largest amount of enhancement due to the contrast media.

T1 and T2 relaxation rates increase with increasing concentration of the paramagnetic substance. At relatively low concentrations, the principle effect seen is that of T1 shortening. With spin-echo T1-weighted images there is an increased signal intensity. The T2 shortening is then insignificant. Increasing concentrations of the paramagnetic continue to shorten T1 and further increase the brightness of the tissue in T1-weighted images — to a certain point. At sufficiently high concentrations, the T2 shortening will

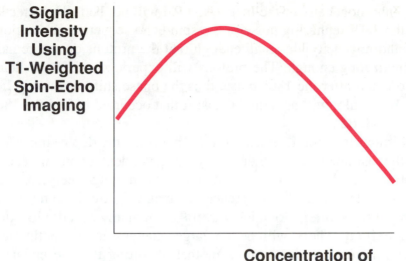

Fig. 16-1. Schematic diagram of the relationship between signal intensity on a T1-weighted spin-echo image and the concentration of a water soluble paramagnetic substance. At low concentrations, the effect of T1 shortening increases the size of the echo signal and thus increases the brightness of the tissue in the image. At higher concentrations, the echo signal is reduced due to the effect of enhanced T2 relaxation which produces significant T2 relaxation during the TE interval. At high enough concentrations the echo signal intensity is dominated by the T2 shortening effect.

become noticeable as a decrease in the signal intensity, even at the short echo times (15-20 ms) used for T1-weighting in the spin-echo or inversion-recovery techniques (Fig. 16-1).

Relatively immobile paramagnetic centers have a mechanism for inducing T2 relaxation effects over a longer range than the few angstroms previously mentioned. If the structure of the ligand containing the paramagnetic center is such that water molecules cannot approach to within a few angstroms of the paramagnetic center, then significant enhancement of the T1 relaxation rate will not occur. However, effective enhancement of T2 relaxation is still possible via the following mechanism. The strong magnetic moment of the relatively immobile paramagnetic center will produce a stable magnetic field inhomogeneity in the region around it. A proton moving (diffusing) near the paramagnetic center will experience a change in the magnetic field strength and therefore a change in its precessional frequency. The dephasing produced by this inhomogeneity of the magnetic field will not be fully corrected by the 180° rephasing pulse of the spin-echo sequence: the regions of inhomogeneity are small enough that the protons do not remain in them long enough. The protons will experience different fields before and after the 180° pulse, thus the phase shift produced after the 180° pulse will not equal the shift that occurred before the 180° pulse (43, 62).

Similar to other T2 relaxation mechanisms, the dephasing effect of the paramagnetic center is highly dependent upon molecular mobility. If the paramagnetic centers are moving freely, e.g., in extracellular fluid, the differences in magnetic field strength that they create will tend to quickly average out at any specific location in the fluid. There will be no large changes in magnetic field strength from one location to another, resulting in little enhancement of T2 relaxation. On the other hand, if the paramagnetic center is immobile, e.g., due to the restraint of a surrounding cell membrane, much larger changes in magnetic field strengths will exist between different locations, resulting in significant enhancement of T2 relaxation (this means a reduction in T2).

Several metal ions, like ferric (Fe^{3+}) and ferrous (Fe^{2+}) iron, manganese (Mn^{2+}) and gadolinium (Gd^{3+}), are strongly paramagnetic, but much too toxic for use as contrast media in their free form. The binding of the metal ion to a carrier molecule (a ligand), thus creating a **chelate**, detoxifies the metal-ion, and may also introduce some organ-specificity for the chelate (63). So far, the only MR contrast medium that has been subjected to extensive clinical trials is gadolinium-DTPA, a chelate of the metal ion Gd^{3+} and the ligand diethylenetriaminepentaacetic acid (63). Gadolinium (Gd^{3+}) has seven unpaired electrons and is the metal ion having the largest magnetic susceptibility. In the currently used dose (0.1 mmol/kg) (41), the effect of Gd-DTPA is primarily that of T1 shortening, i.e., high signal intensity on T1-weighted images, but also some T2 shortening, seen as decreased signal intensity on T2-weighted images (41).

There also exist organic compounds that have unpaired electrons and thus paramagnetic properties. These are called **free radicals**, and they are all unstable with short chemical half lives. Some (e.g., nitroxides) are less unstable than others and are therefore called **stable free radicals**. Although not yet used in patients, these compounds are promising as contrast media. Extensive testing of their safety and efficacy is, however, still needed.

Among the paramagnetic substances normally occurring in the human body, ferrous (Fe^{2+}) and ferric (Fe^{3+}) iron are the most common and widespread. The degradation of hemoglobin that occurs in an intracranial hematoma is a good example of the above-mentioned relaxation mechanisms of paramagnetics. Ferrous iron has four unpaired electrons, but the normally occurring Fe^{2+} deoxyhemoglobin complex has no unpaired electrons and therefore lacks paramagnetic properties (20). In an acute hematoma, this complex releases oxygen and forms Fe^{2+} deoxyhemoglobin which has four unpaired electrons and is paramagnetic (20).

Due to the structure of deoxyhemoglobin, the paramagnetic heme iron is not accessible to water molecules, and T1 relaxation is therefore not enhanced (7, 20). The intracellular Fe^{2+} deoxyhemoglobin has, however, very limited mobility; the strong magnetic moments of the these paramagnetic centers create regions of magnetic nonuniformity. Protons diffusing through these regions will experience relative dephasing, thus T2 relaxation is promoted.

Therefore, on T1-weighted images, an acute (0-2 days) intracranial hematoma appears either equally bright or less bright (due to increased fluid content) compared to brain parenchyma. The hematoma appears less bright on T2-weighted images due to T2 shortening. Since this T2 shortening effect is proportional to the square of the magnetic field strength, it is best seen at high field strengths (4, 20).

In a closed intracranial hematoma, deoxyhemoglobin undergoes oxidative denaturation, forming methemoglobin, during the subacute period from 2 to 14 days (4, 7, 26); if the hematoma is subjected to oxygen (e.g., during surgery), the oxidation may take only a few hours. Methemoglobin contains ferric iron (Fe^{3+}) and is paramagnetic due to the five unpaired electrons of this ion. Ferric iron in methemoglobin is accessible to the water molecules and T1 relaxation is therefore enhanced. A subacute hematoma is consequently bright on a spin-echo T1-weighted image (Fig. 16-2a). The effect on T2 relaxation is dependent upon the mobility of methemoglobin. The formation of methemoglobin precedes lysis of the red blood cells (20), and when methemoglobin is intracellular, T2 is shortened due to the low mobility of the paramagnetic center. After lysis of the red blood cells, the free methemoglobin is highly mobile, and T2 is lengthened (Fig. 16-2b).

The lysed red blood cells are digested by macrophages in the outer rim of the hematoma, and approximately 2-3 weeks after the hemorrhage, hemosiderin is found within lysosomes in these macrophages (20). Hemosiderin contains Fe^{3+}, is water-insoluble (i.e., inaccessible to water molecules) and is very immobile. No T1 shortening but considerable T2 shortening is the result, and the appearance on MR images is that of low signal intensity — most pronounced on T2-weighted images and best seen at high field strengths (Fig. 16-2b).

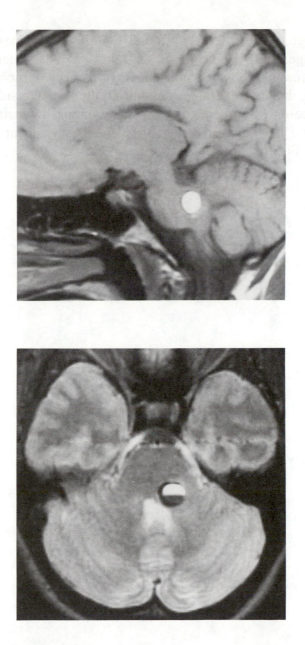

Fig. 16-2. Subacute hematoma in the pons and left cerebellar peduncle. (a) Sagittal T1-weighted image (TR = 600 ms, TE = 20 ms) showing the hematoma as a bright area due to T1 shortening caused by methemoglobin. (b) In this axial T2-weighted image (TR = 2,000 ms, TE = 90 ms) a fluid-fluid level is seen. The supernatant (upper fluid) is bright due to presence of free methemoglobin (long T2); the dependent part of the hematoma (lower fluid) is dark due to T2 shortening caused by methemoglobin within red blood cells. The black rim surrounding the hematoma is due to presence of hemosiderin within macrophages causing T2 shortening.

III. MR Imaging Methods

17. The MR Image, Spatial Information from the MR Signal

In Sections I and II we looked at basic principles of magnetic resonance, production of the MR signal, and mechanisms producing relaxation of the magnetization. We saw how different pulse sequences are used to obtain desired effects on the MR signal intensity due to variations in proton density, T1, and T2. Although we have talked about how different signal intensities are produced by different tissues, we have not yet explained how to differentiate MR signals from various locations in the body.

In a simple, non-imaging, magnetic resonance experiment, a single, net signal is detected from a substance in a sample tube. This net signal is the sum of signals from all parts of the substance. We are not distinguishing among the signals from various parts of the substance; thus we do not obtain spatial information about the substance. The breakthrough in magnetic resonance imaging was the discovery and implementation of techniques to separate the MR signals from different parts of a substance. From a jumble of signals coming from different locations in the body, MRI is able to separate the signals and determine the location from which each originated.

Spatial localization of the MR signals is usually a two step process. First a slice of the body is selected for imaging. A technique is used that allows the protons in only that slice to respond to the applied RF pulse. Protons which respond to an RF pulse are said to be **excited**; the initial 90° or 180° pulse in a pulse sequence is called the **excitation pulse**. The process of interacting with the protons using an RF pulse is called **excitation**.

By exciting only the protons in a slice, the remaining task is reduced to separating the signals from a 2-dimensional slice rather than from a 3-dimensional volume. (Actually, this "slice" has a small thickness so that the MR signal does come from a thin volume. The thickness of this volume is called the **slice thickness** or **slice width**.) Separation of the signals from different locations in the 2-dimensional slice is accomplished by looking at several MR signals from the same slice; each of these signals is produced using a slightly varied pulse sequence. By analyzing and mathematically comparing the differences in these MR signals we can determine the size of the MR signal that comes from each location in the slice.

Each location is then assigned a number which indicates the strength of the MR signal emanating from it. In the MR image these numbers are transformed into shades of grey (or levels of brightness), such that the largest signal produces the lightest shade of grey (or the brightest point) in the image, while the smallest signal produces the darkest grey in the image.

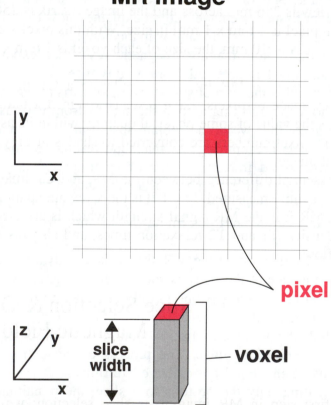

Fig. 17-1. The MR image is divided into a finite number of elements called pixels. Each pixel is an rectangular area in the image. The brightness of a pixel indicates the strength of the MR signal that comes from an associated volume of tissue. This volume is called the voxel. Since the slice of the patient that is represented by the MR image actually has some thickness, called the slice width or slice thickness, the dimensions of the voxel are equal to the pixel's in two dimensions and to the slice width in the third dimension.

In practice, the MR image of a slice is divided into a finite number of elements called **pixels**. A 256 x 256 **image matrix** means that the image is made up of a rectangle that is 256 pixels high and 256 pixels wide, yielding a total of 65,536 pixels (256 x 256 = 65,536). Other common image matrices used in MR imaging are 128 x 128, and 256 x 128. Each pixel in a particular image is assigned a number representing the MR signal intensity from within the pixel; as described above, this number is converted into a density in the image.

Actually, this MR signal comes from a volume of tissue, defined by the pixel in two dimensions and the slice thickness in the third dimension. This volume element is called a **voxel**. Each pixel in the image has an associated voxel (Fig. 17-1). If the image field of view is 256 mm across and the image matrix is 256 x 256, then each pixel is 1 mm x 1 mm in size. With this pixel size, and a slice thickness of 10 mm, the size of each voxel is 1 mm x 1 mm x 10 mm.

In many ways a MR image is similar to a CT image. Both images are made of pixels. Each pixel is assigned a number that represents the value of some physical quantity within its associated voxel. The pixel numbers are converted to shades of grey to form the visual image.

The essential difference between the MR and CT image is the physical quantity measured. In CT it is the x-ray attenuation coefficient; in MR it is the MR signal strength which is affected by the proton density, T1 and T2 relaxation times, and also (as we shall see) by flow.

18. Slice Selection & Orientation, Magnetic Field Gradients

The first step in MR imaging is slice selection: we want to acquire signals from one preselected slice of tissue only. The key to slice selection is the application of **magnetic field gradients**.

Suppose we devise a way to make the static magnetic field increase in strength in a linear fashion in one particular direction. We have then created a magnetic field gradient in that direction. The gradient is a vector that points in the direction of increasing magnetic field strength. The direction of the gradient vector does not depend on the direction of the static magnetic field vector ($\mathbf{B_0}$); these two vectors can point in the same or in different directions. The magnitude of the gradient is determined by the rate of increase of the magnetic field with distance. The magnetic field gradient has units of Tesla per meter (T/m); typical values are 10^{-3} to 10^{-2} T/m (1 to 10 mT/m).

Any plane perpendicular to the gradient direction will have its own unique magnetic field strength. The protons in any such plane will have a common, unique Larmor frequency, different from that

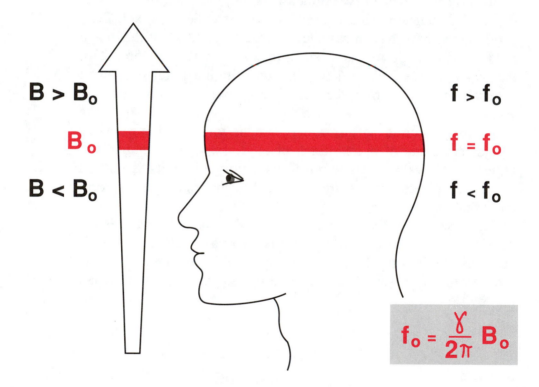

Fig. 18-1. Slice selection by means of a magnetic field gradient. Here the slice selection gradient points in the cranial direction. This means that the strength of the magnetic field increases as we move toward the top of the patient's head. The increasing magnetic field strength is illustrated by the widening arrow in that direction. At a certain plane perpendicular to the gradient direction, i.e., transaxial to the patient's head, the magnetic field due to the gradient will equal zero. The magnetic field in that plane will be simply the static magnetic field, B_0, produced by the MR magnet. All protons in this particular transaxial slice of tissue thus have a Larmor frequency very close to $f_0 = (\gamma/2\pi) B_0$. Protons located cranially to this plane have a higher Larmor frequency; those located caudally to the plane have a lower Larmor frequency. An RF pulse with the frequency equal to f_0 will affect only the protons in the single transaxial slice of tissue where the magnetic field strength is B_0.

of the protons in any other plane perpendicular to the gradient direction (Fig. 18-1). If a large volume of tissue (e.g., the whole head of the patient in Fig. 18-1) is exposed to an RF pulse, this pulse will flip only those protons having a Larmor frequency equal to the frequency of the RF pulse. In the presence of a magnetic field gradient, the flipped protons will be located in one plane only, perpendicular to the gradient direction. This is called **slice selective excitation** of the protons. Provided the protons are not in flowing fluid that leaves the selected slice, the MR signal will be derived only from the selectively excited slice of tissue.

The static magnetic field of the MR magnet has, apart from slight inhomogeneities, the same strength throughout the bore or gap of the magnet. The gradient needed for slice selection is created by specially designed **gradient coils**. There are three pairs of gradient coils located within the bore of the magnet. They are named **x-**, **y-**, and **z-gradient coils** according to the direction of their gradients relative to the static field direction (the z direction).

MR units having a static field oriented along the bore of the magnet will have two circular z-gradient coils, one at each end of the bore of the magnet (Fig. 18-2). The coils are oriented perpendicular to the static field direction. When an electric current runs through a coil, the coil produces a relatively weak magnetic field along the direction of the much stronger static field. A coil's magnetic field can point either parallel or antiparallel to B_0, depending on the direction of its current. (If you curl the fingers of your right hand in the direction of the current, your thumb will point in the general direction of the coil's magnetic field.) Along the axis of a coil, the magnetic field produced by the coil is strongest at the coil's center and decreases as you move in the +z or −z direction away from the center.

Each gradient coil has its own power supply. By running the electric current clockwise in one z-gradient coil and counterclockwise in the other, their magnetic fields will point in opposite directions. In Fig. 18-2, coil no. 1 produces a magnetic field antiparallel to the static magnetic field, B_0. Coil no. 2 produces a magnetic field parallel to the static magnetic field.

The strength of the magnetic field produced by coil no. 1 at its center is b_1; that produced by coil no. 2 is b_2. The magnetic fields b_1 and b_2 are proportional to the electric current in coils no. 1 and 2, respectively. If both coils have the same current, then b_1 will equal b_2. In Fig. 18-2 the net magnetic field is weaker than B_0

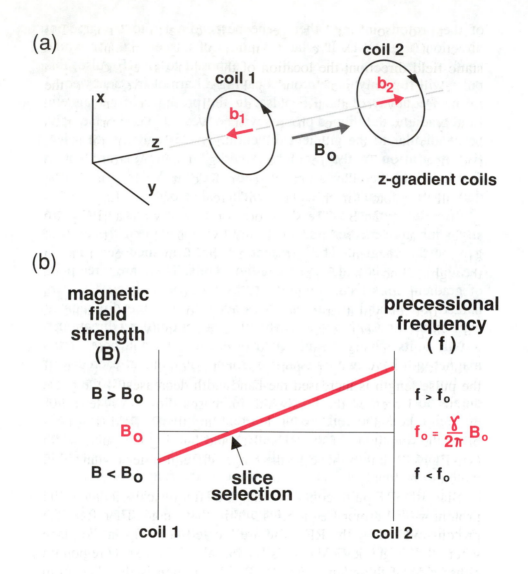

Fig. 18-2. The z-gradient coils and slice selection. Arrows indicate the direction of current flow in the coils that will produce a z-gradient in the positive z direction. (a) The two circular z-gradient coils produce magnetic fields (at their centers) that are antiparallel (b_1) and parallel (b_2) to the static magnetic field, B_0. (b) The combined magnetic fields from the two z-gradient coils result in a magnetic field that varies along the z direction. This magnetic field is weaker near coil 1 and stronger near coil 2. This means that the magnetic field gradient points in the positive z-direction, the same direction as the field due to the MR magnet, B_0. In some plane between the gradient coils the magnetic field from coil 1 and coil 2 will cancel, resulting in a net magnetic field of B_0. This plane, which is oriented perpendicular to the gradient direction, is the selected slice.

close to coil no. 1, and stronger than B_0 close to coil no. 2. The effects of b_1 and b_2 will cancel out in a plane perpendicular to the static field direction; the location of this plane is dependent upon the relative strengths of b_1 and b_2. If the magnetic fields from the two coils have equal strength, they will cancel out in a plane exactly midway between the two coils. When b_1 is stronger than b_2, this plane of cancellation will be located closer to coil no. 1, and vice versa. The plane where the magnetic fields from the two coils cancel out will have a magnetic field strength equal to B_0, and all the protons in this plane will precess with the Larmor frequency $f_0 = (\gamma/2\pi)B_0$. The slice location (sensitive to an RF pulse of frequency f_0) can thus be selected by adjusting the strength of b_1 and b_2 (determined by the strength of the current running through coil no. 1 and 2, respectively).

It is important to note that the RF pulse does not contain only a single frequency; rather it contains a range or band of frequencies. This range of frequencies, also called the **bandwidth** of the RF pulse, can be changed by modifying the length of the pulse. The bandwidth is inversely proportional to the length of the pulse. If the pulse length is increased the bandwidth decreases; if the pulse length is decreased the bandwidth increases. For example, a 90° pulse can be maintained as 90° by doubling the B_1 field (i.e., doubling the amplitude of the RF pulse) and halving the pulse width (see Eq. 6-3), but in this case the range of frequencies contained in the pulse is doubled.

Since the RF pulse contains a range of frequencies, it can excite protons with Larmor frequencies within that range. Therefore, the protons excited by the RF pulse are located not only in the plane where the magnetic field equals B_0, but also in a narrow region on either side of this plane (Fig. 18-3). This region is the slice from which the MR signal will emanate, and its width is the slice thickness. The slice thickness can be increased by decreasing the RF pulse width and thereby increasing its bandwidth; however, it is more commonly controlled by changing the strength of the magnetic field gradient (Fig. 18-3a). Increasing the gradient decreases the distance over which the magnetic field is within the range that can be excited by the RF pulse; thus the slice thickness decreases.

If b_1 and b_2 are changed by the same amount, with the size of one decreased and the other increased, the slice plane can be shifted towards either of the gradient coils without changing the strength of the gradient; the slice thickness will not change (Fig.

18-3b). If both b_1 and b_2 double, the slice plane will not move; however, the magnitude of the gradient will double and the slice thickness will be cut in half.

When the longitudinal axis of the patient is oriented in the static magnetic field direction (the z direction), the use of the z-gradient for slice selection will give slices in the transaxial plane of the patient. (In this case the z-gradient is also called the **slice selection gradient**). If sagittal or coronal slices are desired, the x- or y-gradient coils must be used for slice selection. These paired coils are often saddle-shaped, the long axis of the "saddle" being oriented in the z direction (Fig. 18-4). One of the x-gradient coils produces a magnetic field parallel to the static magnetic field on its side of the saddle "axis," the other a magnetic field antiparallel to the static magnetic field on the other side of the saddle "axis." These two fields thus create a gradient in the x direction and cancel out in a longitudinal plane perpendicular to x, i.e., in a sagittal plane (parallel to the y-z plane) (Fig. 18-4a). Similarly, the y-gradient coils produce a gradient in the y direction, and when used for slice orientation, give a coronal slice parallel to the x-z plane (Fig. 18-4b).

It is important not to confuse the direction of the magnetic field and the direction of the gradient. The z-gradient coils produce a magnetic field along the z direction which varies in strength as z varies. The x- and y-gradient coils also produce a magnetic field along the z direction. However, the magnetic field produced by the x-gradient coils varies in the x direction, and the magnetic field produced by the y-gradient coils varies in the y direction (Fig. 18-5). (Note: the direction of the gradient — or magnetic field variation — may be different from the direction of the magnetic field produced by a pair of gradient coils. The gradient direction and magnetic field direction are different for the x-gradient and also for the y-gradient; the gradient direction and magnetic field direction are the same for the z-gradient coils.)

The combined use of x-, y-, and z-gradient coils can create a magnetic field gradient in any direction. Therefore, with MRI, any slice orientation is possible.

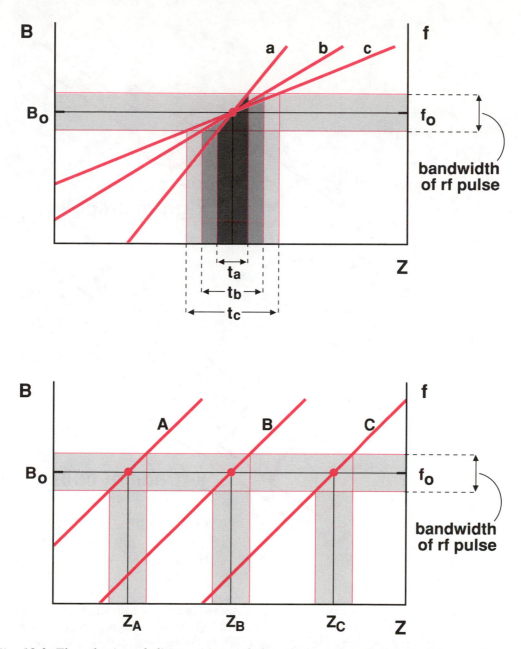

Fig. 18-3. The selection of slice position and slice thickness by adjustment of the gradient strength. (a) The RF pulse which excites the protons contains a small range of frequencies around the desired Larmor frequency. Thus the RF pulse can excite the protons that exist within a small range of magnetic field strength. This range is indicated by the shaded area in the figure. Due to the magnetic field gradient, this range of magnetic field strength will exist within a slice of a particular thickness. The magnetic field strength is plotted as a function of position for three different gradient strengths. These curves are labeled a, b, and c; curve a has the strongest gradient, curve c the weakest. By varying the strength of the gradient, the thickness of the selected slice can be adjusted. Gradient a yields a slice thickness of t_a, gradient b yields a slice thickness of t_b, and gradient c yields a slice thickness of t_c. (b) If the position of the null plane of the gradient is shifted without changing the strength of the gradient, then the position of the selected slice will shift without changing the slice thickness.

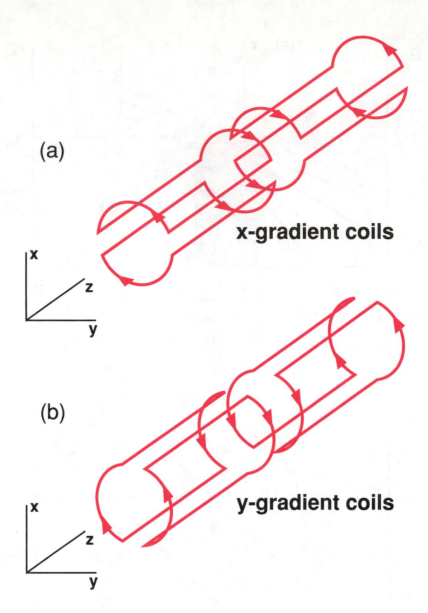

(a)

x-gradient coils

x

z

y

(b)

y-gradient coils

x

z

y

Fig. 18-4. *The x- and y-gradient coils. Arrows indicate the direction of current flow in the coils. Both the x- and y-gradient coils produce magnetic fields within them that point along the z-direction. (a) As drawn here, the x-gradient coils produce a magnetic field that points in the positive z direction near the top of the coils and in the negative z direction near the bottom of the coils. When added to the magnetic field of the MR magnet this causes the total magnetic field to increase in the positive x direction. If all currents in the coils were reversed, the magnetic field would then increase in the negative x direction. (Note that the magnetic field always points in the z direction.) (b). As drawn here, the y-gradient coils produce a magnetic field that points in the positive z direction near the right side of the coils and in the negative z direction near the left side of the coils. When added to the magnetic field of the MR magnet this causes the total magnetic field to increase in the positive y direction. If all currents in the coils were reversed, the magnetic field would then increase in the negative y direction.*

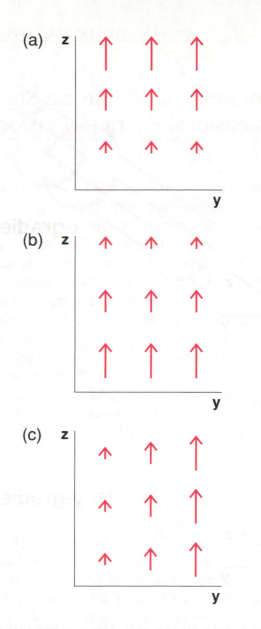

Fig. 18-5. Gradient direction versus magnetic field direction. The arrows represent the net magnetic field vectors at different points in space in the presence of a gradient. The direction of the arrow indicates the direction of the magnetic field vector; the size of the arrow indicates the strength of the magnetic field. (Of course, the magnetic field changes actually produced by the gradients in MR imaging are quite small compared to magnetic field produced by the MR magnet. The changes are greatly exaggerated here for easy visualization.) (a) The magnetic field gradient points in the positive z direction, thus the magnetic field increases in the positive z direction. (b) The magnetic field gradient points in the negative z direction, thus the magnetic field increases in the negative z direction. (c) The magnetic field gradient points in the positive y direction, thus the magnetic field increases in the positive y direction.

Note that in all cases, even though the gradient points in different directions, the magnetic field always points in the positive z direction. The gradient points in the direction you must move to find increasing magnetic field strength. This is a totally separate matter from the direction in which the magnetic field itself points.

19. Spatial Information Within the Slice, Two-Dimensional Fourier Transform (2D FT)

The amplitudes of the signals from a spin-echo, partial-saturation, or inversion-recovery sequence, as described previously, do not contain any information about the spatial location of the signal. To reconstruct an image, we need a method that can separate the MR signals coming from each voxel in the selected slice and then assign those signals to the correct locations in the image, i.e, to the correct pixels. A few methods are available, but the most commonly used technique is **two-dimensional Fourier transform (2D FT)**.

The **Fourier transform** is a mathematical operation that is used in the branch of mathematics called **Fourier analysis**. The basis of Fourier analysis is the following fact: *any* signal can be decomposed (or "analyzed") into a sum of sine waves of different frequencies. Each sine wave can be described by the following expression, in which f is the frequency of the sine wave:

$$A \sin\left(2\pi f t + \phi\right) \qquad (19\text{-}1)$$

Each sine wave at a particular frequency has a characteristic amplitude (A) and a phase (ϕ) (Fig. 19-1). The amplitude is the size of the sine wave. The phase indicates when the zeros and peaks of the sine wave are reached. Changing the phase of a sine wave simply moves the sine wave to the right or to the left along the time axis. The amplitude and phase of each component sine wave is determined by the shape of the original signal.

We touched on the method of Fourier analysis when we discussed magnetic "noise" in tissues (Chap. 9). We talked about separating the noise into its component frequencies; this is a Fourier analysis of the noise.

We are all familiar with one method of performing Fourier analysis: the use of a radio. The air around us is filled with radio waves of many different frequencies coming from a multitude of radio stations. When we tune a radio to a particular station, we are selecting a particular frequency and are detecting the radio waves of that frequency only.

Sin (2 π f t + φ)

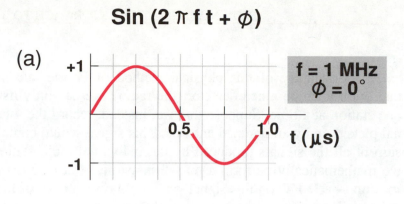

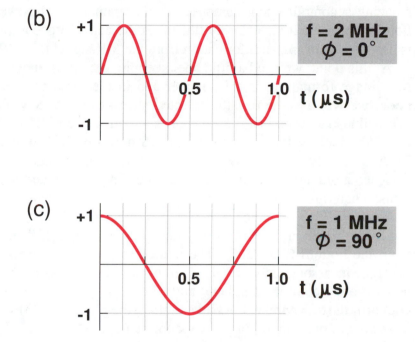

Fig. 19-1. Sine wave graphs illustrating the effects of frequency and phase. (a), (b), and (c) are graphs of the function sin (2πft + φ). A sine wave can be uniquely specified by its amplitude (A), its frequency (f), and its phase (φ). The sine waves shown here all have an amplitude A = 1; however, they differ in frequency and phase. The sine wave is continuous, but only the portion in a 1μs interval is shown in these graphs. (a) and (b) illustrate a change in frequency. In (a) a complete cycle of the sine wave occurs in a time of 1μs (10⁻⁶ sec). The frequency is equal to 1/(10⁻⁶ sec); this is 1 Mhz (10⁶ Hz). In (b) two cycles occur in 1μs, so that the frequency is 2 Mhz or double that in (a). The curve in (c) is identical to that of (a) except that it is shifted to the left by 1/4 of its total cycle length. Since one cycle equals 360°, this curve is shifted by 90°. Its phase (φ) is therefore equal to 90°.

Sine waves can represent many physical phenomena. In this book they will generally represent electrical signals picked up in the receiver coil, i.e., the MR signal. In this case, the horizontal axis indicates time and the vertical axis indicates voltage.

A disadvantage of this physical method of Fourier analysis is that, with a single tuned detector (the radio), we can only listen to one station at a time. Suppose that we, instead, record the total signal picked up by an untuned antenna. This signal would contain the sum of all the signals produced by the individual radio stations. If we mathematically perform a **one-dimensional Fourier transformation — 1D FT**, on this signal, we would obtain a set of individual, separate signals, each with a different frequency and each representing the isolated signal from a single radio station. This amazing accomplishment demonstrates the mathematical power of Fourier analysis. It also closely illustrates the actual procedure that is used to obtain spatial information from the MR signal (Fig. 19-2).

At this point, we need a little bookkeeping to help us understand MR image formation. In this chapter and in the remainder of the book, we will assume that the selected slice lies in the x-y plane. We will also assume that our image has a matrix size of 256 x 128, i.e., 256 pixels in the x direction and 128 pixels in the y direction. Each of the 256 x 128 or 32,768 pixels in the image can be assigned a unique location or address according to its x and y positions. There are different, equivalent ways we can do this. In this book we will label the columns or x positions from 1 to 256 starting from the left and the rows or y positions from 1 to 128 starting from the bottom (Fig. 19-3). The unique location or address of a pixel (or its associated voxel) is given by stating its x position or **x-coordinate** (column number) and then its y position or **y-coordinate** (row number). As an example, the pixel (74, 23) would have an x-coordinate of 74 and a y-coordinate of 23.

Our task in producing an MR image is to separate the signals that come from each of the 32,768 voxels comprising the image. Suppose that we apply an x-gradient while detecting the MR signal (FID or echo). Each voxel with a different x-coordinate will then be in a slightly different magnetic field. As a result the protons (and the magnetization vectors) in these voxels will each be precessing with slightly different frequencies; the MR signals coming from voxels with differing x-coordinates will have different frequencies. By applying a 1D FT to the composite MR signal, we can separate out the signals originating at different x locations. This helps us somewhat, but we still have a problem in that each of these component signals remains the sum of signals from all the voxels in a column, i.e., those with the same x-coordinate but different y-coordinates. We have successfully separated out the x-

Fourier Analysis

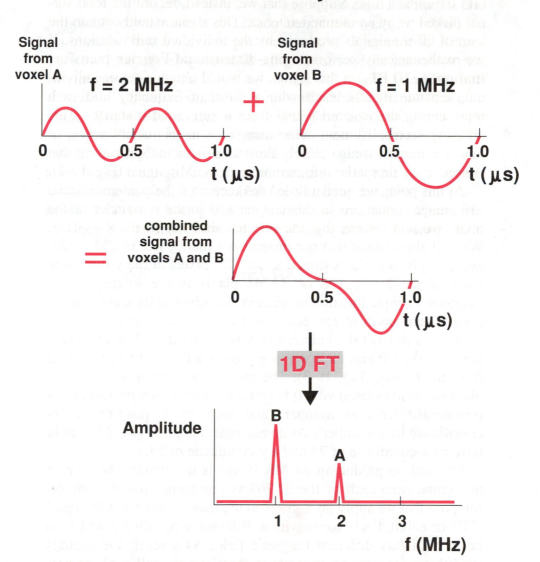

Fig. 19-2. The combination and analysis of sine waves. If sine waves of different frequencies are added together, the result is a complex looking waveform that is no longer a simple sine wave. Any conceivable signal can be created by adding together sine waves of different frequencies. Logically then, any signal can be decomposed into its component sine waves. This mathematical process is called one-dimensional Fourier transformation (1D FT). When a 1D FT is performed on a signal, the result is a set of numbers which indicate both the amplitude and phase of each of the component sine waves of different frequency. In this figure the signal is the sum of two sine waves of frequency 1 MHz and 2MHz. A 1D FT of the signal then indicates that it is made up of sine waves of 2 frequencies and gives their amplitudes. Although not shown in this figure, the phases of the sine waves would also be given by the 1D FT. (Here, both sine waves have a phase equal to zero.)

dimension but not the y-dimension. No matter how we try to orient or modify the gradient we cannot overcome this problem using the 1D FT with a single MR signal.

So far we have separated out the x-coordinates by encoding each x-coordinate with a unique frequency. We will separate out the y-coordinates by encoding the different y-coordinates with different phases (Fig. 19-4). Suppose that, for a brief moment, we apply a y-gradient after the first pulse that excites the protons in the slice but *before*, not *during*, the detection of the MR signal. (This is most easily done when using an echo as the MR signal for constructing an image, since there is a reasonable time interval between the first RF pulse and the echo.) During the application of the y-gradient, the precessional frequency of the protons and the magnetization in a voxel varies according to the y-coordinate of that voxel. When the y-gradient is turned off, the protons and the

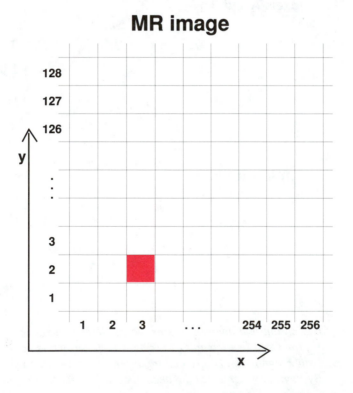

Fig. 19-3. Pixel coordinates in the MR image. Each pixel in the image can be given a unique address according to its location in the image. This figure shows a 256 x 128 image matrix. The pixel indicated in red has the coordinates (3, 2) or x = 3 and y = 2.

magnetization vectors again precess with (approximately) the same frequency in every voxel, but the protons and the magnetization in each voxel now have an additional phase shift. The amount of phase shift is determined by the y-coordinate of the voxel. In a voxel that is in a higher magnetic field during the y-gradient, the protons and the magnetization precess with a higher frequency when the y-gradient is on; they also rotate through a larger total angle and accumulate a larger phase angle compared to those that precess at lower frequencies. Thus, after the application of the y-gradient pulse, there will be a linear relationship between phase angle and location in the y direction.

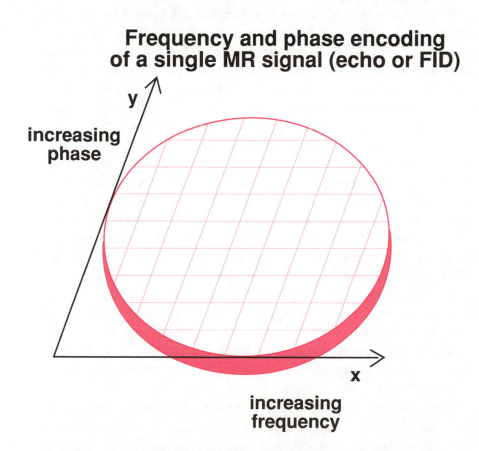

Frequency and phase encoding
of a single MR signal (echo or FID)

Fig. 19-4. Frequency and phase encoding. This is a schematic drawing of a slice of tissue indicating the individual voxels. The application of the x-gradient while detecting the MR signal assigns a unique frequency to the voxels according to their x-coordinates. The application of a y-gradient for a short period before the detection of the MR signal assigns a varying phase to the voxels according to their y-coordinate.

When using a single MR signal with the properly applied x- and y-gradients, it would be nice if we could separate the signals coming from voxels with different y-coordinates by looking at their phases, in the same way we separated the signals from voxels with different x-coordinates by looking at their frequencies. However, this is not possible.

When sine waves of different frequencies are added together, the result is a complex looking signal that may not even remotely resemble any of its component sine waves. However, it is always possible to analyze this composite signal and decompose it into the original sine waves, as long as each component sine wave is a different frequency (Fig. 19-2).

When sine waves of different phases but the same frequency are added together, the result is another sine wave of the same frequency but with a phase and amplitude dependent on the amplitudes and phases of the component sine waves* (Fig. 19-5). After these sine waves are added together there is no way to decompose the result into the original sine waves. (Because the composite signal is a pure sine wave it cannot be further decomposed.)

Although we cannot separate the signals coming from different y-coordinates based on the phase encoding of a single MR signal, we can separate these signals if we obtain several MR signals, each with a different phase encoding. This is accomplished by varying the size or the time duration of the y-gradient in each successive pulse sequence. To perform the separation of MR signals originating from 128 different y positions we need to obtain 128 separate MR signals. Each of these 128 MR signals must be produced using a pulse sequence containing a y-gradient with a different size or time duration. Each MR signal obtained using a different y-gradient is called a **view**. (Similarly, to separate the MR signals from 256 different x-coordinates we must be able to adequately separate each MR signal into 256 sine wave components of different frequency.)

The most common method of producing an MR image uses the

---

* The addition of sine waves of the same frequency can be reduced to a problem in vector addition. In this method, each sine wave is represented by a vector whose size is the amplitude of the sine wave and whose direction is determined by the phase angle of the sine wave (the vector is oriented so that it makes an angle with the x-axis equal to the phase angle of the sine wave.) The vector representing the composite sine wave is simply the sum of the vectors representing the component sine waves.

spin-echo pulse sequence. In Fig. 19-6 several successive pulse sequences are shown, each consisting of one 90° pulse and one 180° echo pulse. Before the very first 90° pulse is transmitted (this first pulse is not shown in Fig. 19-6), all tissues are fully magnetized; i.e., their protons are all at thermal equilibrium, with the magnetization vector oriented in the static magnetic field direction at its maximum size. Thus the echo amplitude in this first pulse sequence is usually larger than those in the subsequent pulse sequences, and therefore it is excluded from the imaging process.

In the example below, slice selection is determined by the z-gradient. The application of the z-gradient during each transmission of the 90° and 180° pulses assures that the received echoes are derived only from protons excited within the selected slice of tissue. An RF pulse that affects only the protons within a selected slice is called a **selective pulse**. Such a pulse must have a restricted bandwidth and be applied during a slice selection gradient so that it only affects the protons within the selected slice thickness. A **non-selective** RF pulse affects the protons throughout an entire volume of tissue; its effect is not restricted to a single slice. To apply a non-selective RF pulse the slice selection gradient is either kept off during the pulse; or the duration of the RF pulse is greatly shortened, and its bandwidth consequently increased, so that it can affect a large thickness of tissue even in the presence of a gradient. If a selective 90° pulse is used, the 180° does not actually need to be selective to assure proper imaging of the desired slice. However, if the 180° pulse is non-selective, it will affect the protons throughout the tissue volume; the magnetization in tissues outside the selected slice will be inverted by the 180° pulse. This will have no effect on the image of a single slice. However, we shall see in the next chapter that it is possible to obtain images of several slices simultaneously. In this imaging mode, the inversion of magnetization in tissues outside of a slice interferes with the imaging of other slices. Thus the 90° and 180° pulses are normally both selective pulses.

When the z-gradient is used for slice selection, the x- and y-gradients produce magnetic field variations in the plane of the selected slice. The y-gradient is turned on for a brief moment, after the z-gradient has been turned off and before the echo is induced. As discussed previously, this produces a phase shift in the angle of the precessing protons and magnetization in each voxel; this phase shift varies with the y position of the voxel. Immediately after the y-gradient is turned off there will be a linear relationship between

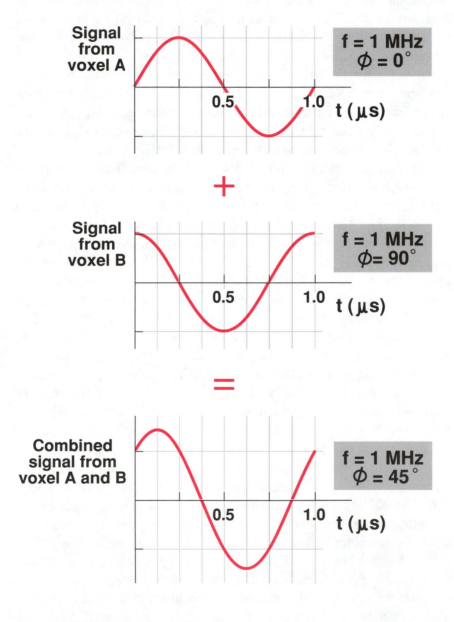

Fig. 19-5 (facing page). The addition of sine waves with the same frequency, but different phases. When sine waves with the same frequency but different phase are added together, the result is not a complex waveform but a simple sine wave. Once the sine waves are added together there is no way to analyze the result and regain the original sine waves. Thus, while it is possible to separate out the frequency encoding of a single MR signal it is not possible to separate out the phase encoding. To perform this last feat we need multiple MR signals each with a different degree of phase encoding.

Suppose we added the two sine waves together a second time; in this second addition we shift the phase of the "voxel B" sine wave by some known amount. If we compare the results of the first and second addition, it is now possible to determine the phases and amplitudes of the original two sine waves. If we wanted to determine the phases and amplitudes of 100 different sine waves of the same frequency after they had been added, we would have to perform 100 separate additions. In each individual addition every sine wave would be given a different multiple of some known phase shift before being added. For each of the 100 additions this phase shift would be different.

the phase angle, produced by the y-gradient, and the y-coordinate of the voxel.

After the y-gradient is turned off, the 180° echo pulse is applied at the desired time (its timing after the 90° pulse is one-half the selected TE interval). After the 180° pulse, the x-gradient is turned on. The x-gradient remains on for the entire duration of the echo signal. Thus, during the echo signal, the precessional frequency of the magnetization in each voxel is determined by the x-coordinate of the voxel. (Since the x-gradient is on while the echo is being detected or "read," it is also called the **readout gradient,** and the x direction is often termed the **readout direction.**)

To produce a single 256 x 128 image we must perform 128 of these 90-180 pulse sequences, each yielding an echo. The following is a concise, chronological listing of the events in an individual pulse sequence: (1) z-gradient pulse and 90° pulse together, (2) y-gradient pulse, (3) z-gradient pulse and 180° echo pulse, (4) x-gradient pulse during readout of echo signal.

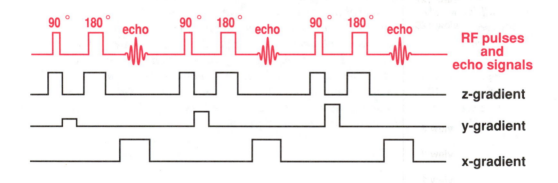

Fig. 19-6. Basic spin-echo pulse sequence in MR imaging using the 2D FT method. In this and similar succeeding diagrams the RF pulses and MR signals appear together in red on the top line (or lines). Though all the RF and gradient pulses are shown schematically as turning sharply on and off, in reality the transitions are a bit more gradual. In this diagram, 3 cycles of the pulse sequence are shown. They are all identical except for the changes in the phase-encoding y-gradient. The slice is selected by application of the z-gradient during transmission of the RF pulses. The phase encoding y-gradient is increased step-wise for each pulse sequence. The frequency encoding x-gradient is applied throughout the detection of the echo. Note: The rephasing portions of the z- and x-gradients have not been drawn here (see Fig. 19-8). This figure and similar ones appearing later are not drawn to scale: the time between the echo and the next RF pulse is drawn much smaller than normal. This is done so that multiple cycles of the pulse sequence can be shown. In general, TR is much longer than TE, although it does not appear to be so in these figures.

The only variation in the applied RF pulses and gradient pulses among the 128 pulse sequences is with the y-gradient. One method of variation of the y-gradient is to use a negative gradient in the first pulse sequence that is strong enough so that the phase of each successive voxel in the y direction is changed by 180° relative to its two neighbors with the same x-coordinate. The second pulse sequence will use a y-gradient that is 63/64 the strength of the first y-gradient. The third pulse sequence will use a y-gradient that is 62/64 the strength of the first y-gradient, and so on. In each successive pulse sequence the y-gradient will change by 1/64 of the original gradient strength. The y-gradient in the 64th pulse sequence will be nearly zero and later pulse sequences will use positive pulse gradients that successively increase in strength until the last gradient is the same strength as the first (but opposite in direction).

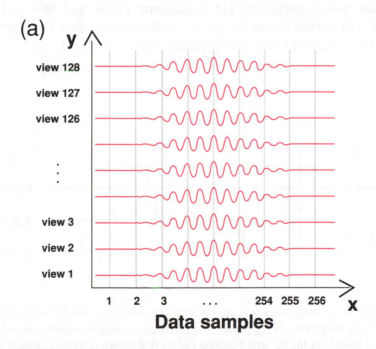

Fig. 19-7. Image formation using the 2D FT method. (a) To reconstruct an MR image on a 256 x 128 matrix, 128 echo signals must be obtained, each from a separate pulse sequence having a different y-gradient strength. Each signal is sampled at 256 equally spaced times. In this figure, each of these separate MR signals or "views" is plotted on an x-y coordinate system. An individual view is oriented so that time runs along the x-axis. Each view is positioned in the y-direction according to the phase encoding gradient used. (In our example, the view with the largest negative phase encoding gradient is view 1 and the view with the largest positive phase encoding gradient is view 128.) (explanation continues on next page)

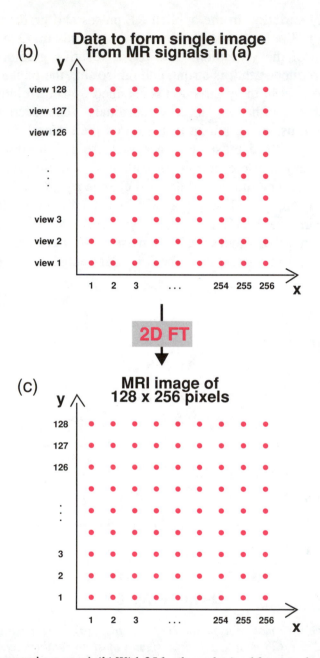

Data to form single image from MR signals in (a)

(b)

MRI image of 128 x 256 pixels

(c)

2D FT

(cont. from previous page) *(b) With 256 values obtained from each of 128 separate views, this figure schematically shows each of these 256 x 128 or 32,768 data points as a dot on the x-y plot. If a 2D FT is taken of this 2-dimensional set of data points, the result is another 2-dimensional set of data points. (c) These latter data points comprise the 2-dimensional pixel matrix of the MR image. Each dot represents a separate pixel value in the MR image. Each pixel value is proportional to the MR signal that originated from within the voxel of tissue associated with the image pixel.*

As each echo is "read out" it is sampled at 256 points in time (Fig. 19-7a). For the moment let's look at the echo obtained with a pulse sequence having zero y-gradient. If a 1D FT is performed on the 256 data points from this echo, the result will be 256 amplitudes and phases, each representing a sine wave of a different frequency. Each of these sine waves is the sum of the MR signals from all voxels having the same x-coordinate (and therefore producing the same frequency MR signal).

Let's now look at the echo data points from all 128 echoes (Fig. 19-7b). Each echo has 256 data points; this number times 128 echoes yields a total of 32,768 data points (the same as the number of pixels). We can plot these points so that we form a 256 x 128 matrix of data points: the 256 data points from an echo are plotted horizontally, along the x direction; each echo's data points are plotted so that their y-coordinate is proportional to the y-gradient used to produce that echo. If a 2D FT is now taken of this data matrix, the direct result is the image matrix of 256 x 128 pixels (Fig. 19-7c).

The above considerations should make it clear why the pulse sequences, whether saturation-recovery, spin-echo, or inversion-recovery, need to be repeated many times in MR imaging. The number of repetitions depends upon the number of pixels in the phase encoding (y) direction. If 128 pixels are wanted in the y direction, 128 repetitions of the pulse sequence are needed; for 256 pixels, 256 repetitions must be performed. This is the minimum required number of repetitions.

To increase the signal-to-noise ratio, each pulse sequence can be repeated one or more times without changing the y-gradient. The number of pulse sequences preformed with the same y- (and z- and x-) gradient is called the **number of excitations (NEX)**. If an image with 2 NEX is wanted, the y-gradient is increased one step with every other pulse sequence; if 4 NEX is needed to obtain the desired signal-to-noise ratio, the y-gradient is increased one step with every fourth pulse sequence. In that case, the pulse sequence will need to be repeated 4 x 128 (or 512) times.

Due to the large number of repetitions, the acquisition of an MR image can be a lengthy procedure. The acquisition time is equal to the product of the repetition time (TR), the number of excitations (NEX), and the number of phase encoded "views", i.e., the number of pixels in the y direction. A typical imaging time for a proton density or T2-weighted image, both of which have a long TR, is thus: 2,000 ms (TR) x 2 (NEX) x 256 (phase encoded "views") =

17.1 min — and this is for just one image. The mathematical computations needed for reconstruction of the image take only a few seconds. Clearly, an imaging time of 17.1 min per image is unacceptable. The acquisition of 15-20 T2-weighted images (a common number of images needed) would then take approximately 5 hours. Fortunately, there are methods to dramatically decrease the acquisition time of conventional MR images. These methods will be discussed in the next chapter.

Note that the above discussion is a simple overview of the MR imaging process. Various fine points were omitted so that we could concentrate on more essential matters. We will now discuss the basic features of some of these refinements.

The selective 90° pulse used to excite the protons is not simply switched on and off. We want the 90° pulse to contain a small range of frequencies, but we would also like it to contain an equal amount of each frequency within this range. This is achieved by giving the amplitude or "envelope" of the RF pulse a particular shape called the **sinc function**. In this way the RF signals from the tissues near the boundary of the slice thickness will not be appreciably weakened, i.e., the slice sensitivity will be approximately equal throughout the selected slice thickness.

We have seen how to produce intentional phase shifts by the use of the y-gradient. Unfortunately, the use of the x- and z-gradients also produce phase shifts by the same mechanisms as those produced by the y-gradient. These unwanted phase shifts require corrections to the pulse sequence to remove their effects.

As the magnetization in a voxel is rotated from the z direction to the x-y plane by the 90° pulse, the protons with different z locations within the voxel will precess with slightly different frequencies and will acquire varying phase shifts due to the slice selection gradient. We can compensate for these unwanted phase shifts using a process called **gradient reversal** or **gradient rephasing**. Immediately after a z-gradient is applied for slice selection, another z-gradient is applied but in the opposite direction at about half-strength (or half-duration) (Fig. 19-8). This negative z-gradient does not affect the slice selection, since the 90° pulse has already ended when the negative gradient begins. However, it does compensate for the proton dephasing produced by the positive z-gradient.

The x-gradient also produces proton dephasing which would destroy the echo signal unless a "pretreatment" is applied. This pretreatment involves applying an x-gradient (again at about half

the strength or duration of the later x-gradient pulse) during the
period in which the y-gradient is applied (Fig. 19-8). This pretreatment x-gradient does not need to be reversed in sign: since it
occurs before the 180° echo pulse, the 180° pulse effectively
reverses the gradient's effect on proton dephasing. This x-gradient
(before the 180° pulse) dephases the protons with different x-
coordinates. When the 180° pulse is applied the relative phases of
the protons are inverted: after the 180° pulse the protons that were
ahead in phase will now lag. The application of an x-gradient (in
the same direction as the first x-gradient) after the 180° pulse will
cause the protons to come back into phase. The timing and strength
of this x-gradient must be adjusted so that the time of the spin-echo
produced by the 180° pulse coincides with the rephasing of the
protons due to the x-gradient. If the x-gradients before and after the
180° pulse are the same strength, then rephasing of the protons will
occur after the second x-gradient pulse (readout pulse) has been on
a time equal to the duration of the first x-gradient pulse. Since this

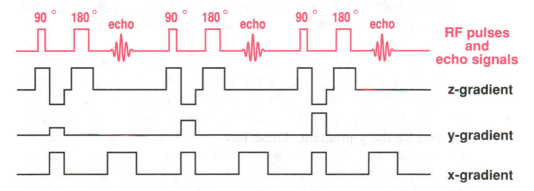

*Fig. 19-8. Basic spin-echo pulse sequence in MR imaging using the 2D FT method including
the required time-reversal gradients. This figure is identical to Fig. 19-6 except for some
additions to the sequence of z- and x-gradients. A negative z-gradient is added after the positive slice selection z-gradient, compensating for the unwanted phase shift introduced by the
positive gradient. (As an exercise you may wish to consider that this negative z-gradient could
be eliminated by simply lengthening the time of the second positive z-gradient so that its
length after the 180° pulse is greater than its length before the 180° pulse. This extension of
the second positive z-gradient will have the same rephasing effects as a negative gradient
placed between the 90° and 180° pulse.) If the readout gradient is applied as indicated in Fig.
19-6, it will dephase the protons and destroy the MR signal that is expected at the time of the
spin-echo. To correct this problem and insure that the protons are not dephased by the x-
gradient at the time of the echo, a positive x-gradient must be applied after the 90° pulse to
initially dephase the protons. The subsequent readout x-gradient will then rephase the protons. (Both gradients can be positive, due to the phase inversion properties of the 180° pulse
that exists between the gradient pulses.)*

rephasing should occur at the midpoint of the readout pulse, the readout pulse will be twice the length of the first x-gradient pulse.

The z-gradient that is applied during the 180° pulse does not require any additional "correction" gradient pulses as long as the z-gradient has equal duration before and after the 180° pulse. In this case the phase shifts that are produced during the first half of the z-gradient pulse are corrected during the second half of the z-gradient pulse due to the phase inversion produced by the 180° pulse.

The extra gradient pulses that are added to compensate for unwanted dephasing are sometimes called **time-reversal gradients**. They are essential to the imaging process.

In Section II we discussed transverse relaxation and the difference between T2 and T2* relaxation. T2* relaxation occurred more rapidly than T2 relaxation because of the additional proton dephasing due to the magnetic field inhomogeneities produced by the MR magnet. Both the FID and the echo signals decreased with the T2* time constant. In an imaging pulse sequence, the MR signal (FID or echo) decreases even more rapidly because of the readout gradient (x-gradient), which is an added intentional magnetic field inhomogeneity.

What would happen if we removed the 180° pulse from the pulse sequence; could we obtain a degree of "refocusing" of the protons and produce a type of echo using only the gradients? The answer is yes. If the 180° pulse is removed from our "standard" pulse sequence, an echo can still be produced near the middle of the x-gradient pulse. This is called a **gradient echo**. A gradient echo differs from an echo produced using a 180° pulse in that the amplitude of a 180° echo depends on the T2 relaxation time in the tissue while the amplitude of the gradient echo depends on the T2* relaxation time and is thus smaller. To produce this gradient echo, the time-reversal x-gradient must be opposite in sign from the later "refocusing" x-gradient (here there is no 180° pulse to reverse the effect of the first x-gradient). This first x-gradient causes dephasing of the protons. The later x-gradient then reverses this dephasing, thereby "refocusing" the protons and producing a gradient echo (Fig. 19-9).

In Section II we discussed several pulse sequences that use the FID as the MR signal. In MR imaging the existence of the FID immediately after the 90° pulse poses severe problems: it is in practice difficult to measure the first part of the FID that occurs immediately after the 90° pulse. There also would seem to be no

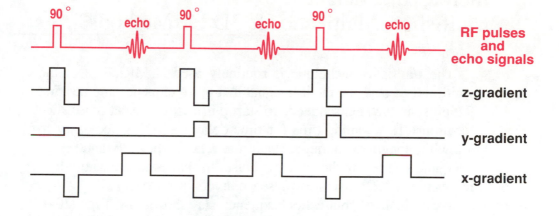

Fig. 19-9. Basic gradient echo pulse sequence. This sequence is very similar to Fig. 19-8, except that the 180° echo pulse is missing, and the first x-gradient pulse is opposite in sign from the readout x-gradient pulse. These two x-gradient pulses respectively dephase and rephase the protons to provide an in-phase condition and thus an MR signal or echo at the desired time after the 90° pulse. The absence of the 180° pulse means that dephasing due to magnetic field inhomogeneities is not cancelled out at the time of the echo. This reduces the size of the echo; in this case the echo size is the result of T2* relaxation subsequent to the 90° pulse rather than T2 relaxation.

time period in which to place the y-gradient and the time reversed x- and z-gradients before the MR signal begins. These difficulties can be overcome either by changing to an equivalent pulse sequence that uses a 180° echo pulse (e.g., in Chapter 15 we discussed the addition of a 180° echo pulse to an inversion-recovery pulse sequence) or by effectively "postponing" the onset of the FID by the use of time-reversal gradients. A time-reversal x-gradient applied immediately after the 90° pulse will abort the FID; if the readout x-gradient is then applied quickly thereafter, a gradient echo, little affected by T2* relaxation, will be produced a short time after the 90° pulse.** This gradient echo is then like a slightly delayed FID except that it provides the bonus of a rising and falling side; it is like two FID curves back to back. The additional information provided by the rising part of the echo is quite useful in the image reconstruction process.

The topics of gradient reversal and gradient echoes will appear again in Chapters 21 and 22.

** This sequence of events is identical to that shown in Fig. 19-9, except that the positive x-gradient is applied immediately after the negative x-gradient, thus shortening the TE interval.

20. Increased Efficiency: Multislice, Multiecho, & 3D FT Methods

The **multislice** technique is routinely used in MRI. It uses the waiting time between the last signal in a pulse sequence and the first RF pulse in the next sequence to start pulse sequences at other slice locations. This waiting time (or time delay — TD) is necessary to allow the longitudinal magnetization to relax sufficiently and regain adequate size before the next 90° pulse (in spin-echo and saturation-recovery) or 180° inversion pulse (in inversion-recovery).

A multislice spin-echo sequence is shown in Fig. 20-1. Immediately after the registration of the echo from the first slice (z_1), the slice selective z-gradient is shifted to another slice location (z_2) and a second complete set of RF and gradient pulses are transmitted. When the echo from the second slice is registered, the z-gradient is shifted to the third slice location (z_3), and so on. In this way, images from several slice locations can be obtained in the same time interval that we previously calculated was required to produce a single image.

The number of slices that can be excited during the period of one time delay naturally depends upon both the echo time (TE) and the time delay (TD). With a longer time delay (and thus longer repetition time — TR) and/or a shorter echo time, more slices can fit into the time delay available. The maximum number of slices that can be obtained with the multislice technique can be determined from the selected values of TE and TR:

$$\text{Maximum number of slices} \quad = \frac{\text{TR}}{(\text{TE} + \text{C})} \qquad (20\text{-}1)$$

This equation expresses the number of pulse sequences that will "fit" within the time TR. The effective duration of the pulse sequence is somewhat greater than TE, because data acquisition of the echo signal continues after the echo peak occurs at the time TE, and also because of normal software and hardware limitations. This additional time is represented by the constant, C, in Eq. 20-1 and is about 10 to 20 msec. Using a spin echo pulse sequence with TE = 80 msec and TR = 2000 msec (SE 2000/80) we could obtain about 20 slices simultaneously using multislice imaging. Instead of

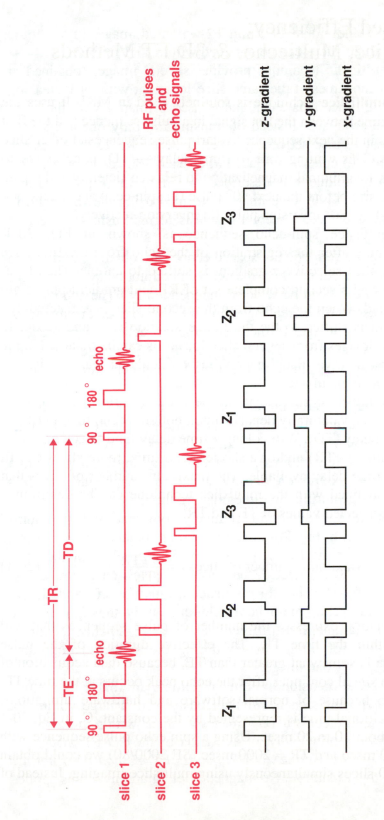
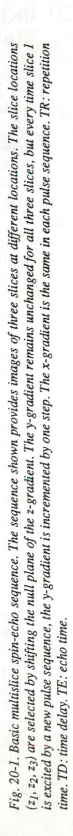

Fig. 20-1. Basic multislice spin-echo sequence. The sequence shown provides images of three slices at different locations. The slice locations (z_1, z_2, z_3) are selected by shifting the null plane of the z-gradient. The y-gradient remains unchanged for all three slices, but every time slice 1 is excited by a new pulse sequence, the y-gradient is incremented by one step. The x-gradient is the same in each pulse sequence. TR: repetition time. TD: time delay. TE: echo time.

obtaining 20 individual 17.1 min T2-weighted images in 5.7 hours (Ch. 19), all 20 slices can be obtained within 17.1 minutes.

The **multiecho** technique provides several images, obtained at different echo times at the same slice location, without increasing the imaging time. This technique is used primarily with the spin-echo pulse sequence. Instead of transmitting only one 180° echo pulse in each pulse sequence, several successive 180° pulses are transmitted. An echo is produced after each 180° echo pulse. This multiecho technique was described in Chap. 13 (Fig. 13-3) and Chap. 14 (Fig. 14-1). One image is formed using the first echo in each pulse sequence; a second image is formed using the second echo, and so on.

In Fig. 20-2, four equally spaced echoes are obtained during one pulse sequence. (In practice, the echoes need not be evenly spaced; TE for each echo can be selected independently.) The echo amplitude decreases with increasing echo time and so does the signal-to-noise ratio. The images formed using the last echoes therefore tend to be somewhat noisy. The term "heavily T2-weighted" is often used for images obtained with especially long echo times (at least 80-90 ms).

Even though a large number of echoes may be obtained during the pulse sequence, two are usually sufficient for diagnostic purposes. The use of a long TR (e.g., 2,000 ms) and two echoes, one with a short TE (e.g., 20 ms), the other with a long TE (e.g., 90 ms) will provide one proton density weighted image and one T2-weighted image during the same imaging time.

The multiecho technique is routinely combined with the multislice technique. In Fig. 20-2 there appears to be an extremely short time delay after the last echo, but that is normally not the case. When a long TR is selected, there usually is sufficient time delay, even after a fourth echo, for multislice acquisition of 15-20 slices. Using a multiecho, multislice technique, it is thus possible to obtain, for example, four different echoes at 15-20 slice locations, giving a total of 60-80 images during the same imaging time required for only one image.

The **three-dimensional Fourier transform (3D FT)** method involves acquiring signals simultaneously from a large volume of tissue and then reconstructing images of slices within that volume. This method is principally an extension of the 2D FT method to three dimensions. Unlike 2D FT, *the 3D FT method does not use slice selective excitation*. No magnetic field gradient is thus applied

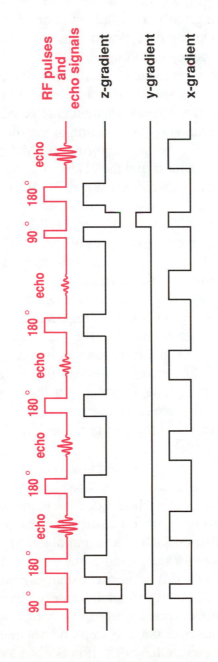

Fig. 20-2. Basic multiecho sequence. This figure shows a spin-echo sequence producing four echoes in each se-quence. The y-gradient is increased one step for each new pulse sequence and is applied only once per sequence. The x-gradient is applied during the detection of each echo signal. In this figure there appears to be little time be-tween the last echo and the next 90° pulse, in this sense the figure is not to proper scale. Usually the time between the last echo and the next 90° pulse is much longer than the time between the previous 90° pulse and the last echo. This allows pulse sequences from other slices to be inserted within this waiting time. Thus the multiecho technique can be combined with the multislice technique.

during RF pulse transmission. This means that all the protons in the large volume of tissue are flipped by the RF pulses.

Immediately after the first RF pulse in each pulse sequence, both y- and z-gradients are applied; these gradients produce phase shifts of the protons according to their locations in the y and z directions. (In 2D FT imaging, only a y-gradient is applied at this point for phase encoding.) During detection of the echo signal, an x-gradient is applied for frequency encoding.

With 3D FT, images can be computed directly with any orientation of the image slice using the acquired MR signals. This contrasts with 2D FT imaging in which the slice orientation is selected before data collection by the determination of the direction of the slice selection gradient. In 2D FT imaging, before acquiring the MR signals, we must also select the slice thickness and the 2D pixel matrix in the x and y directions (e.g., 10 mm and 256 x 128). In 3D FT imaging, before acquiring the MR signals, we must select the 3D voxel matrix in the x, y, and z directions (e.g., 256 x 128 x 64). If a 256 x 128 x 64 matrix were selected, we could produce 64 slices of 256 x 128 matrix images in the x-y plane or 128 slices of 256 x 64 matrix images in the x-z plane, as two possibilities for slice orientation.

In 2D FT imaging, the pulse sequence is repeated many times with different values of the y-gradient. The required number of repetitions is equal to the number of pixels in the y direction of the image. In 3D FT imaging, the pulse sequences must also be repeated many times, using different values of the y- and z-gradient. The required number of separate values for the y-gradient is still equal to the number of voxels in the y direction. The z-gradient must also have a number of separate values equal to the number of voxels in the z direction.

During the phase encoding portion of the pulse sequence, the y- and z-gradients are applied simultaneously. Their values can be adjusted according to the following method. For each required value of the z-gradient, pulse sequences are obtained using each of the required values of the y-gradient. (Each MR signal obtained using a separate combination of values for the y- and z-gradients is referred to as a "view.") To accomplish this task, a constant z-gradient is selected and the pulse sequences are repeated with ever increasing y-gradient. When all the phase encoded views in the y direction are obtained, the z-gradient is incremented one step and another complete set of phase encoded views in the y direction is obtained. This process continues until all steps of the z-gradient have been used.

The minimum number of pulse sequences required to obtain the complete y and z phase encoding is equal to the number of voxels in the y direction times the number of voxels in the z direction. The imaging time will be equal to the product of following four factors: TR, NEX, the number of phase encoded views in the y direction, and the number of phase encoded views in the z direction.

Full use of the 3D FT method would allow reconstruction of images in any plane with equally high resolution. This would require the same high number of pixels (128 or 256) in all directions (x, y and z), and would lead to a very long imaging time (hours) when combined with conventional pulse sequences like saturation-recovery, spin-echo and inversion-recovery. Often, to save time, images in only one plane are reconstructed. If axial slices, i.e., slices perpendicular to the z direction, are selected for reconstruction, then the number of phase encoded views in the z direction can be reduced to the number of slices needed (e.g., 10-30). The pixel size in the z direction is adjusted to the desired size of the slice thickness.

3D FT reconstruction has a potential advantage over the 2D FT method in terms of the time required to perform a study. With 3D FT, each MR signal is derived from an entire volume, rather than from a single slice; this allows simultaneous reconstruction of multiple slices. This advantage is mostly overcome by the use of the multislice technique with 2D FT imaging; in this way the 2D FT method can also produce simultaneous reconstruction of multiple slices. However, we shall see in the next chapter that multislice imaging cannot normally be used with fast imaging methods that use very short repetition times, excitation RF pulses of less than 90°, and gradient echoes. With these fast imaging methods, 3D FT reconstruction can be a very efficient imaging technique with substantial time advantages over 2D FT reconstruction (15, 30).

21. Fast Imaging:
Small Flip Angles & Gradient Echoes

Multislice, multiecho, and 3D FT imaging techniques can dramatically reduce the time required to complete an MRI examination by obtaining many images simultaneously. The production of an MR image using the conventional pulse sequences described so far is a fairly slow process. To prevent imaging artifacts, patients must hold perfectly still for several minutes at a time, and imaging of small children usually requires sedation.

There are several possible ways to decrease imaging time (15, 21, 27, 39, 42, 56, 57). Currently the most widespread fast imaging technique uses short repetition times, excitation RF pulses of less than 90°, and gradient echoes. There are several variants of this fast scanning technique, and each has been given a special name by the group that developed the variant. The variants include CE-FAST (Picker), DEFT (Picker, Philips), FISP (Siemens), FLASH (Max Planck, Gottingen), and GRASS (General Electric). The pulse sequence described below is the GRASS variant, but the basic principles are applicable to other fast-scan techniques that use a short TR, **flip angles*** less than 90°, and gradient echoes. The decrease in the imaging time is the direct result of the decrease in TR; as will be explained, the small flip angles and the gradient echoes allow the use of short TRs. While the above imaging techniques reduce the data collection time for an image, they also can provide additional useful contrast possibilities, including sensitivity to fluid flow and chemical shift effects, and are sometimes used to take advantage of these image contrast features.

Consider a saturation-recovery or spin-echo pulse sequence. As TR is shortened, the longitudinal magnetization is allowed less time to return to its maximum (equilibrium) amplitude. Thus the size of the longitudinal magnetization just prior to the 90° pulse is reduced; and the subsequent MR signal (FID or echo) has a decreased amplitude. A very short TR (much shorter than T1) produces an MR signal too weak to be useful for imaging. When the TR is this short, using a small angle RF pulse (less than 90°), instead of a 90° pulse, will increase the MR signal intensity (Fig. 21-1).

---

* The flip angle is the angle of the RF excitation pulse, the first pulse of the pulse sequence. It is called the flip angle since it is this RF pulse that rotates, or flips, the longitudinal magnetization from its orientation along the z-axis. All pulse sequences discussed in previous chapters had flip angles of 90° or 180°.

An excitation RF pulse with a small angle has the advantage of producing a significant amount of transverse magnetization (M_{xy}), while retaining much of the original, positive longitudinal magnetization (M_z). For example, a 45° pulse produces an M_{xy} that is 71%

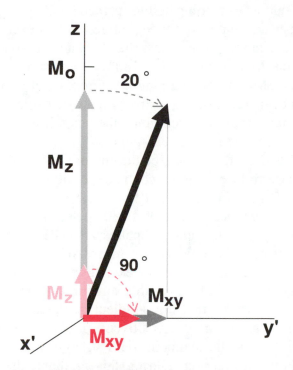

Fig. 21-1. Effect of flip angle on signal intensity at very short TRs (TR << T1). This figure shows the magnetization vectors just before and just after a second RF excitation pulse using flip angles of 90° (red) and 20° (black). After an initial 90° pulse, the longitudinal magnetization (M_z) grows from zero toward its maximum, equilibrium value M_0. If TR is very short, little time is allowed for M_z to increase before the next 90° pulse is applied. Thus M_z will be much reduced in size (light red vector) just prior to this second 90° pulse. The transverse magnetization (M_{xy}) that results (dark red vector) after the second 90° pulse will also be greatly reduced in size, thus the MR signal will be weak. After an initial 20° pulse, the longitudinal magnetization (M_z) is still large, fully 94% of its equilibrium value, M_0. Even if TR is very short, so that M_z only grows a minimal amount before the next 20° pulse is applied, M_z will still be quite large (light grey vector) just prior to this second 20° pulse. The magnetization (black vector) after the second 20° pulse will be equally large. The transverse component of this magnetization, M_{xy}, (medium grey vector) will be only 34% of the size of the total magnetization vector. Even so it may be substantially larger than the M_{xy} that results from the second 90° pulse. Thus, when using very short TR, RF excitation pulses with flip angle of less than 90° can substantially increase the strength of the MR signal.

of the original M_z; immediately after this RF pulse, M_z retains 71% of its original amplitude.** Immediately after a 20° pulse, M_{xy} equals 34% of the original M_z, while M_z retains 94% of its original amplitude! After a small flip angle pulse, M_z is much closer to its equilibrium value than it would be after a 90° pulse; thus TD (and TR) can be quite short, compared to methods using 90° pulses, and still achieve an adequate size for M_z before the next excitation pulse.

After an excitation pulse of less than 90°, a positive longitudinal magnetization (M_z) is present. If a 180° echo pulse were then used, it would invert M_z, thereby moving it far from equilibrium and nullifying the advantage of using a small flip angle excitation pulse. It is therefore inappropriate to combine the very short TR, small flip angle pulse sequence with a 180° pulse. Still, the technical advantages of an echo are desirable (see the end of Chap. 19) and the use of an echo introduces added contrast possibilities due to T2* relaxation effects.† In Chapter 19 we saw that it is possible to produce echoes without the use of a 180° pulse. These echoes, produced using magnetic field gradient pulses, are called gradient echoes.

A gradient echo is obtained by means of a gradient reversal (Fig. 21-2). Immediately after an excitation pulse of angle α, a time-reversal x-gradient (the readout direction) is applied; for this discussion let's assume that this gradient is negative. Though the magnetic field points in the positive z direction, a negative x-gradient means that the magnetic field strength increases linearly in the negative x direction. During this gradient, the protons in each voxel precess with a frequency that depends on the x-coordinate of the voxel. The x-gradient is then reversed, thus creating increasing magnetic field strength in the positive x direction. The protons that had the fastest precession during the negative gradient now have the slowest precession, and vice versa. The effect of the negative gradient is a dephasing of the protons; that of

---

** If an RF pulse of angle α is applied to a tissue with a magnetization, M_0, in the z-direction, the resulting values of longitudinal and transverse magnetization are given by the following equations:

$$M_z = M_0 \cos (\alpha)$$
$$M_{xy} = M_0 \sin (\alpha)$$

† In this chapter we will talk about T2* relaxation effects since, strictly speaking, with the absence of a 180° echo pulse, the size of the gradient echo depends on T2* relaxation rather than simply T2 relaxation. For a particular MR scanner, however, we expect that T2* would change in response to the T2 variations in the tissues. A T2*-weighted image should reflect the T2 variations in the tissue.

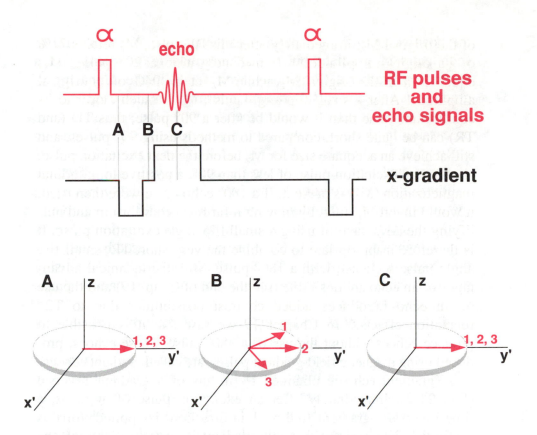

Fig. 21-2. *Proton excitation using a small flip angle RF pulse and formation of the gradient echo. When an RF pulse of angle α, less than 90°, is used to excite the protons, it is no longer practical to use a 180° pulse in the echo formation process. In this case the echo must be generated solely by the x-gradient.*

The upper part of the figure: After the application of a small flip angle RF pulse, a negative x-gradient is applied to dephase the protons. The positive x-gradient applied during readout of the RF signal then rephases the protons and generates the gradient echo. If both gradients have the same strength, the maximum amplitude of the echo will occur when the positive gradient has been on as long as the negative gradient. The positive gradient must continue after this point so that it is on during detection of the entire echo. This provides the necessary frequency encoding of the signal.

The lower part of the figure: The transverse component of the magnetic moment vectors of three protons (1, 2 and 3) located within the same voxel, but at slightly different locations in the x direction. The rotating frame of reference is used. The frame of reference is rotating with the precessional frequency of proton 2. (A) It is assumed that the protons are precessing in phase prior to the application of the first x-gradient. During the negative x-gradient the three protons will precess with slightly different frequencies — proton 3 with the highest and proton 1 with the lowest. (B) The three protons after the end of the negative x-gradient pulse. Proton 3 has precessed the farthest and is ahead of protons 2 and 1. During the positive x-gradient proton 1 precesses with the highest frequency and proton 3 with the lowest. (C) At the time of the echo, protons 1 and 2 have caught up with proton 3, so that all three protons are now in phase.

the positive gradient, a rephasing of the protons. We can refer to such pairs of gradient pulses as a **dephasing gradient** and a **rephasing gradient**. (Previously, we referred to the dephasing x-gradient as a time-reversal gradient, a term that is also used.)

Prior to the negative gradient the protons of the spin excess are precessing in phase with the same precessional frequency, and an MR signal briefly exists (A in Fig. 21-2). (Let's ignore the effects of the y- and z-gradients for the moment; they do not affect the validity of our explanation.) This MR signal is quickly destroyed by the x-gradient as the protons are dephased. After the negative x-gradient at point B in Fig. 21-2, the protons have experienced maximum dephasing; from B to C the protons are rephased by the positive x-gradient. At point C the effects of the two opposing gradients cancel out; here, the protons are once more back in phase and produce an echo.

If the x-gradient were the only cause of the proton dephasing from A to C, then M_{xy} at C would be the same as M_{xy} at A; however, there are other processes causing dephasing of the protons: T2 relaxation and the magnetic field inhomogeneities of the MR magnet. These processes by themselves would cause M_{xy} to decrease with a relaxation time of T2*. In other words, if left alone (not disturbed by the gradients), the protons would have induced an FID after the α pulse in Fig. 21-2, and the amplitude of the FID would have decayed with the time constant T2*. By dephasing the protons the negative x-gradient quickly "destroys" the FID. The positive x-gradient cancels out the effect of the negative x-gradient and brings back a signal (the echo) that has the same amplitude that the FID would have had at that time if no gradients had been applied. The amplitudes of successive gradient echoes after an α pulse decrease with the time constant T2* (Fig. 21-3), instead of T2 (which is the case with echoes produced by 180° pulses). The process of creating gradient echoes does not cancel out the effects of magnetic field inhomogeneities; the production of echoes with 180° pulses does cancel these inhomogeneities. Images obtained from gradient echoes are therefore much more susceptible to loss of signal and degradation of image quality due to magnetic field inhomogeneities than are those obtained from spin echoes.

Fig. 21-4 provides a summary of the basic sequence of RF and gradient pulses used in the fast imaging technique. A comparison of this figure with Fig. 19-8 will show that it is similar to the basic spin-echo pulse sequence except for 3 differences: (1) the excitation pulse is an α pulse instead of a 90° pulse; (2) the 180° echo pulse and its associated slice selection gradient is missing; and (3) the dephasing (or time reversal) x-gradient that is applied after the excitation pulse is negative rather than positive. (In the spin-echo pulse sequence, the 180° pulse effectively inverts the effect of this dephasing gradient pulse.)

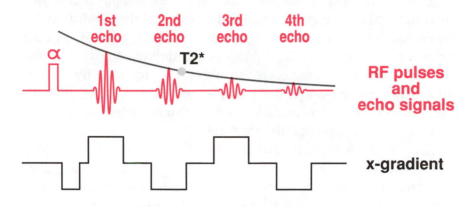

Fig. 21-3. Multiple gradient echoes after a single RF pulse. The maximum amplitude of the echoes decays approximately exponentially with the time constant T2*. The part of the readout gradient after the peak of the first echo continues to dephase the protons as it provides the required frequency encoding to the MR signal. This part of the readout gradient is itself the dephasing gradient for the next echo signal. To rephase the protons and form a second echo, a negative x-gradient must be applied during readout of the second echo. Thus multiple gradient echoes can be formed by using readout gradient pulses that change sign for each successive echo.

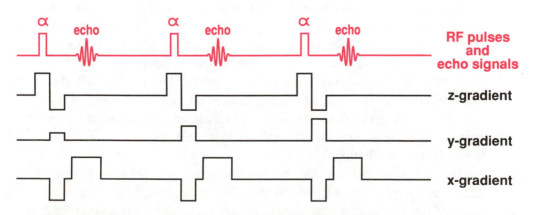

Fig. 21-4. Basic gradient echo pulse sequence using a small flip angle RF excitation pulse. This sequence is essentially identical to Fig. 19-9 except that the 90° pulse has been replaced with an α pulse.

It may seem interesting that the spin-echo pulse sequence is essentially just a gradient echo pulse sequence with an added 180° pulse. In fact, the spin-echo sequence does contain a gradient echo. It occurs simultaneously with the echo due to the 180° pulse and allows this echo to manifest itself.

Saying that the gradient echo and the echo due to the 180° pulse occur simultaneously does not mean that there are physically two superimposed echo signals. Rather, it means that the conditions for the production of an echo due to gradient rephasing and due to the rephasing effects of the 180° pulse occur simultaneously. To distinguish it from the gradient echo, the echo produced by a 180° pulse is sometimes called the **Hahn echo**. We have seen that the Hahn echo occurs at a time after the 180° pulse equal to the time between the 90° pulse and the 180° pulse.

Since a readout gradient is required in the pulse sequence to provide frequency encoding, the pulse sequence must always contain a pair of dephasing-rephasing gradient pulses and a resulting gradient echo. An MR signal can exist only around the time of the gradient echo, when the effects of the dephasing and rephasing gradients cancel out. Thus the existence of a gradient echo at a point in time is a *requirement* for the MR signal to exist at that time.

The echo produced by the 180° pulse, on the other hand, is actually only an enhancement to the gradient echo: by compensating for the effects of the dephasing due to magnetic field inhomogeneities, the Hahn echo increases the strength of the gradient echo, but a Hahn echo is *not required* to have an MR signal. In the absence of a Hahn echo, the gradient echo can still occur; however, if the first dephasing x-gradient is left out of a spin-echo pulse sequence so that the gradient echo does not occur, then no MR signal will appear at the time of the Hahn echo. If the timing of the 180° pulse is changed so that the Hahn echo in a spin-echo pulse sequence does not occur simultaneously with the gradient echo, the MR signal will occur at the time of the gradient echo, not at the time of the Hahn echo. Slight offsetting of the gradient and Hahn echo times can actually have some useful effects in chemical shift imaging as we shall see in Chapter 23.

Fig. 21-4 shows us that no time-reversal (or rephasing) gradients are applied along the y-axis. That is logical, since the purpose of the y-gradient is to create phase shifts along the y-axis that are still present when the echo is induced. Naturally, we do not want the phase shifts in the y direction to be transferred into the next pulse sequence. In conventional pulse sequences this is taken care of by the relatively long TRs. All protons are completely dephased long before the start of the next pulse sequence. However, when using a fast imaging technique with extremely short TRs, there still remains some transverse magnetization when the next RF pulse is transmitted. Phase encoding in the y direction will therefore still be present at this time. When using this fast imaging technique, an additional y-gradient pulse must be added in the time interval between the gradient echo and the next RF pulse to cancel the remaining phase shift (56, 57).

In Fig. 21-4 we see that each pulse sequence in the fast imaging technique consists of one RF pulse of less than 90° followed by one gradient echo, the amplitude of which is dependent upon T2*. Image reconstruction can be done either by the 2D FT or the 3D FT method. With fast imaging techniques using very short TR (and TD), multislice imaging is not normally available; thus, 3D FT imaging can provide significant advantages in reducing the total examination time. With fast imaging modes using moderate TR times (i.e., 200-600 ms), multislice imaging is possible and is now available with some MRI scanners. The contrast in images is dependent upon three adjustable parameters: repetition time (TR), echo time (TE) and flip angle (α). Their effect on contrast is rather complex, as the effect of one parameter is dependent upon the values of the others (25, 32, 33, 56, 57, 58).

We shall first explore the effect of the flip angle on image contrast when TR is extremely short (20-50 ms). Naturally, TE must be even shorter (8-15 ms). A TR this short will be close to or even shorter than the T2* of many tissues. This means that when the next RF pulse is transmitted, the dephasing of the protons is not complete; there is still a magnetization component in the transverse plane, at least in some of the tissues. The effects of a relatively large flip angle ($\alpha = 30°\text{-}60°$) under these circumstances are shown in Fig. 21-5.

Consider a collection of protons at equilibrium with a magnetization vector (M_0) along the z-axis. The magnetization is then rotated α degrees by an RF pulse (Fig. 21-5a). After completion of the RF pulse, T1 and T2* relaxation effects begin to increase the size of M_z and reduce the size of M_{xy} respectively. If T2* is extremely short, M_{xy} will nearly disappear before the next RF pulse is applied (Fig. 21-5b). The magnetization vector is now oriented nearly along the z-axis and is considerably shorter than

M_0 due to the relatively large flip angle and incomplete T1 relaxation during TR. (T1 will be significantly longer than TR, so that only a little T1 relaxation occurs in a time TR). The next RF pulse rotates the magnetization α degrees (Fig. 21-5c), after which T1 and T2* relaxation again occur. Each time the magnetization is rotated by the α pulse, M_z is reduced to a fraction of its size just before the α pulse; this fraction is given by cos α. (Since cos 45°= 0.71, a 45° pulse will reduce M_z to 71% of its original value.)

After several α pulses, the value of M_z just before the RF pulses will stabilize at a value that may be less than 20% of M_0 (Fig. 21-5e). Stabilization at a small value of M_z can be explained in the following way. When M_z is small it is farther from its equilibrium value of M_0; therefore M_z will change more rapidly with time (Fig. 8-2). Furthermore, the rotation of a small M_z by the RF pulse causes only a small absolute change in M_z (the rotation reduces M_z by a fixed percentage, but as M_z decreases, this fixed percentage of a smaller M_z yields a smaller absolute change in M_z). The faster change in M_z produced by T1 relaxation coupled with the smaller change in M_z that needs to be recovered in order to return M_z to its size before the RF pulse allows stabilization of M_z to occur at small values of M_z, when it could not occur at larger values. Because M_z stabilizes at a small value, the M_{xy} produced by the α RF pulse will also be small (it will be sin α times M_z).

If, on the other hand, T2* is long (longer than TR), significant M_{xy} will still exist when the next α pulse is transmitted (Fig. 21-5b). The size of the magnetization vector just before the RF pulse will therefore be larger than that in the case of the short T2* and will also be oriented at a fairly large angle to the z-axis (although less than the angle α due to some T1 and T2* relaxation).

Let us suppose that the direction of $\mathbf{B_1}$ is inverted for every other RF pulse. This means that if the magnetization is rotated clockwise by the first α pulse, it will be rotated counter-clockwise by the second pulse, clockwise by the third, and so on. After being rotated counter-clockwise α degrees by the second RF pulse, the magnetization will make an angle of less than α with the z-axis (Fig. 21-5c).

As a result of this smaller angle and the larger size of the magnetization, M_z will be considerably larger than in the corresponding short T2* case. After several RF pulses the magnetization will also stabilize, as in the short T2* case, but at a significantly larger size (Fig. 21-5e). The angle of the magnetization relative to the z-

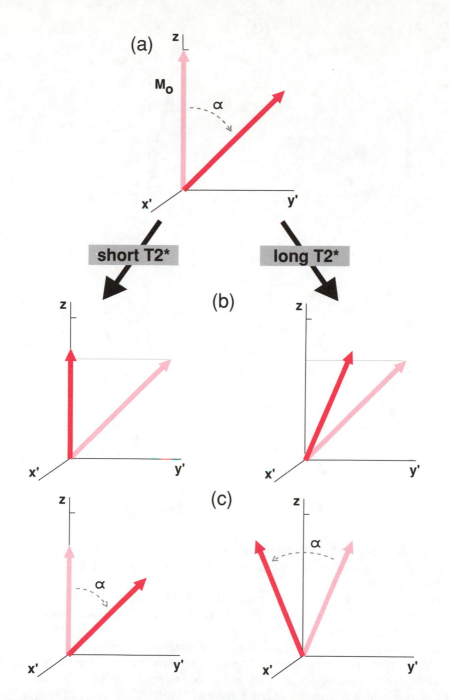

Fig. 21-5. Effect of T2* on MR signal intensity using a relatively large flip angle (30°- 60°), an extremely short TR (20-50 ms) and a very short TE (8-15 ms). All figures are depicted in the rotating frame of reference. (a) A collection of protons at equilibrium. Their magnetization vector, M_0, is rotated α degrees by an RF pulse. (b) The magnetization vector just after the first RF pulse is shown in light red, the magnetization vector just before the second RF pulse is shown in dark red. The change in the magnetization vector is due to T1 and T2* relaxation. (c) The rotation of the magnetization vector by the second RF pulse. The magnetization is indicated by light red before the RF pulse and dark red after the pulse. (d) The magnetization vector just after the second RF pulse is shown in light red, the magnetization vector just before the third RF pulse is shown in dark red. (explanation continues on next page)

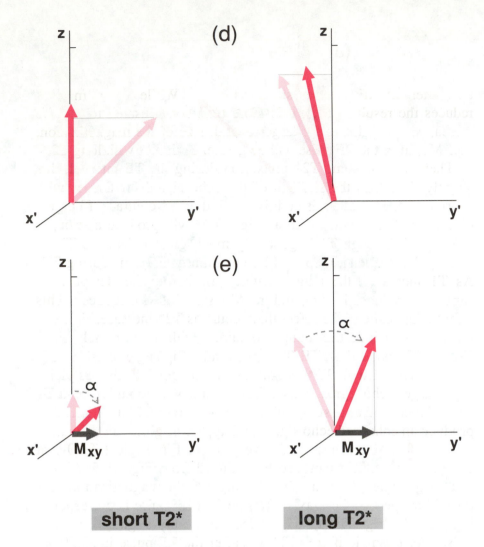

(d)

(e)

short T2*   **long T2***

(cont. from previous page) *(e) The magnetization vector just before (light red) and just after (dark red) the RF pulse — after the application of several RF pulses. This indicates the value at which the magnetization stabilizes. Also shown (in grey) is the transverse magnetization, M_{xy}, produced immediately after the RF pulse.* **Short T2*:** *(b) Since T2* is short, complete loss of transverse magnetization occurs before the next RF pulse is applied. Since T1 is usually much longer than T2 and T2*, only incomplete T1 relaxation occurs before the next RF pulse. Thus the magnetization is oriented along the z direction, but with a size substantially less than M_0. (c) After the second RF pulse the magnetization is oriented at an angle α with the z direction. (d) Just before the third RF pulse the magnetization again points in the z-direction but with a further reduction in size compared to (b). (e) Due to a substantial loss of longitudinal magnetization after several RF pulses, the M_{xy} that results after an RF pulse is also small.* **Long T2*:** *(b) Since T2* is long, some transverse magnetization remains just before the next RF pulse is applied. Thus, the magnetization vector is oriented at an angle to the z direction. This remaining transverse magnetization also increases the size of the magnetization vector. (c) After the second RF pulse, the magnetization is oriented at an angle to the z-axis of less than α, so that a large fraction of the longitudinal magnetization is retained after this pulse. (d) Relaxation of the magnetization after the second RF pulse leaves a fairly large magnetization vector. (e) After several RF pulses, the magnetization stabilizes at an orientation of less than α degrees with the z-axis and with a much larger size than in the case for short T2*. Because of the larger magnetization vector, the M_{xy} that results after an RF pulse is larger than in the short T2* case. Thus the MR signal is larger with longer T2*.*

axis after an RF pulse is smaller than α. While this somewhat reduces the resultant M_{xy}, it is more than compensated for by the increase in M_{xy} due to the larger overall size of the magnetization, thus M_{xy} after the RF pulse will be greater in tissues with long T2*.

There will be some T2* relaxation during the TE interval; this slightly decreases the strength of the echo signal for tissues with shorter T2*. This effect is additive with the T2* contrast effect just discussed. Thus tissues with a longer T2* will produce a stronger MR signal. Though T1 is generally much greater than the TR of these pulse sequences, some T1 relaxation does occur during TR. As T1 increases, the magnetization just before the RF pulse is somewhat reduced in size, and the MR signal is also reduced. This effect is opposite to the effect that occurs as T2* increases. We can therefore conclude that images obtained with an extremely short TR (20-50 ms), a short TE (8-15 ms) and a flip angle of 30° to 60° are, to a large extent T2*-weighted, with longer T2* tissues yielding a larger echo signal. This effect is somewhat modified by a T1 dependence so that tissues with a higher ratio of T2* to T1 tend to produce an enhanced echo signal and appear brighter in the image.

The effect on contrast of a very small flip angle (5°-20°) at extremely short TRs can also be deduced from Fig. 21-5. A very small flip angle (α) means that the angle of the magnetization vector with respect to the z-axis after the first RF pulse is correspondingly small (Fig. 21-5a). M_z is decreased only slightly by the α pulse. For example if $\alpha = 10°$, M_z after the RF pulse is 98.5% of M_z before the RF pulse (cos 10° = 0.985). Therefore, M_z will stabilize at a relatively large value even if T2* is very short. As with larger flip angles, the stabilization value of the magnetization will be larger as T2* increases, but in this case, the differences will be small. As a result, the M_{xy} produced by the RF pulse will only be affected slightly by T2* variations. Due to the very short TE (8-15 ms) the effects of T2* relaxation during TE period are minimal. Thus the amplitude of the gradient echoes will depend only weakly on T2*. T1 effects on echo strength are also minor: due to the small change in M_z produced by the RF pulse, T1 relaxation can produce only a small change in the size of the magnetization. Images obtained with an extremely short TR and a very small flip angle are therefore largely proton density weighted. (Remember that all MR images are affected in the same way by proton density variations: the MR signal is always proportional to the proton density. However, since T1 and T2 effects can easily dominate the image contrast, an imaging mode is called "proton density weighted" only if the T1 and T2 effects are slight.)

Next we will explore the effect of the flip angle on contrast when the TR is only moderately short (200-600 ms) and TE is very short (8-15 ms). TR is now considerably longer than the longest T2*, and M_{xy} is zero when the RF pulses are transmitted. Furthermore, TE is so short that only a small amount of T2* decay occurs during TE, thus T2* plays only a minor role in determination of contrast. Under these circumstances, image contrast is largely dependent upon T1 and flip angle. Here TR is long enough that a significant degree of T1 relaxation can occur during the TR interval.

If the flip angle is small (5°-20°), M_z is only slightly reduced by the application of the RF pulse (as discussed previously). As before, with small flip angles we find that the size of the magnetization before the RF pulse will stabilize at relatively large values. In this case, its stabilization value will be slightly larger for tissues with shorter T1s, but the differences will be slight. For this pulse sequence T1 and T2* dependence is minor, thus image contrast is principally controlled by proton density.

If the flip angle is larger (45°-90°), M_z is substantially reduced by the application of the RF pulse. If the T1 relaxation rate is high (short T1), then T1 relaxation will be able to return M_z most of the way to M_0 by the end of the TR interval. In this case, the magnetization before the RF pulse will stabilize at a relatively large value. M_{xy} after the RF pulse will be corresponding large and so will the echo signal. However, if the T1 relaxation rate is low (long T1), much less relaxation will occur during the TR interval. Thus the magnetization before the RF pulse will stabilize at a much smaller value; M_{xy} after the RF pulse will also be small and so will the echo signal. The use of a relatively large flip angle produces images that depend mostly on T1 variations for image contrast. Tissues with shorter T1 will provide a strong MR signal and appear bright in the image.

In summary, images obtained with a TR significantly longer than the longest T2* (i.e., TR = 200-600 ms), a short TE (8-15 ms), and a small flip angle (5°-20°) are primarily proton density weighted. When the flip angle is increased to 45°-90°, the images will be largely T1-weighted. It is generally true that a decrease in the flip angle decreases the amount of T1-weighting.

So far, in our discussion of gradient echo pulse sequences that use moderate TRs of 200-600 ms, we have considered only very short TEs of 8-15 ms. Using a longer TE would increase T2*-weighting. When TE is increased, differences in echo amplitude due to differences in T2* are amplified; tissues with short T2* will suffer a greater loss of echo signal than tissues with longer T2*. If we start with a proton density weighted pulse sequence that uses a very small flip angle, increasing TE will change it into a T2*-weighted pulse sequence. If we try to introduce T2*-weighting into a T1-weighted pulse sequence that uses larger flip angle we run into difficulties. In this case T1 and T2* relaxation have opposing effects on the image contrast: longer T1 produces image darkening while longer T2* produces image brightening. We could encounter the same unfortunate loss of image contrast that we saw was possible with other standard pulse sequences. Thus when imaging with increased TE, the flip angle should be kept small to minimize the T1-weighting. Any T1-weighting would counteract the T2-weighting and decrease the image contrast.

Typical parameters for a heavily T2*-weighted image are a fairly long TR of 200-600 ms, a long TE of 30-60 ms, and a small flip angle of 5°-20°. However, as previously discussed, a substantial (although not "heavy") T2*-weighting can be obtained with extremely short TRs when these are combined with fairly large flip angles (30°-60°) and short TEs (8-15 ms), though this pulse sequence is also affected by the ratio T2*/T1.

It is probably now quite clear that the causes of image contrast due to fast imaging techniques are fairly complex. Not all possible parameter combinations have been explored above, but some of the most useful ones have been mentioned. Two aspects of image contrast have yet to be discussed. This fast imaging technique is very susceptible to flow related enhancement, a phenomenon discussed in the next chapter. Furthermore, due to the gradient echoes, chemical shift can have a peculiar effect on image contrast. This will be dealt with in Chapter 23.

22. The Signal Intensity of Flowing Blood & CSF

Unlike CT and other imaging methods using x-rays, MRI is able to image the lumen of blood vessels without the use of contrast media. The difference in MR signal intensity from the blood within the vessels compared to surrounding tissue is due to the fact that the blood is flowing, rather than due to the relaxation characteristics of the blood itself. Due to flow, the lumen of blood vessels can appear either very bright or very dark compared to the surrounding tissue. Among the many factors affecting the MR signal intensity from blood are the blood flow velocity, the velocity profile (e.g., parabolic or "plug" flow), the flow characteristic (laminar or turbulent) the type of pulse sequence used and its time intervals, the strength and orientation of the gradients, the orientation of the slices with respect to the vessels, the use of single-slice or multi-slice technique, and the image reconstruction method. The situation is obviously quite complex.

The flow velocity of arterial blood varies during the cardiac cycle. The highest velocity will be found during systole, the lowest velocity during diastole. Thus the timing of the MR pulse sequence might be expected to affect the appearance of arterial blood in the image.

Normally, each pulse sequence used to form a particular image will start at a different part of the cardiac cycle. The appearance of the lumen of arteries in such an image will be the sum of effects produced by the different velocities during the cardiac cycle. However, it is possible to have each pulse sequence begin at the same point of the cardiac cycle if a technique called **EKG-gating** is used. In this technique the fist RF pulse transmission in each pulse sequence is triggered by the R-wave (i.e., the largest peak) in the EKG. The TR of the pulse sequence thus becomes determined by the R-R interval, i.e., the time period from one heartbeat to the next. If we allow each R-wave to trigger the transmission of the 90° pulse in the spin-echo pulse sequence, the TR of that pulse sequence becomes equal to the R-R interval. If we allow only every other R-wave to trigger the 90° pulse (which is no problem technically), the TR of the pulse sequence becomes equal to twice the R-R interval.

Multislice imaging is possible with EKG-gating. In this case, the excitation of the first slice is triggered by the R-wave, and the next slices are "filled in" during the time delay period (TD) until the next R-wave induced excitation of the first slice. The data from each slice will thus be collected from different locations within the cardiac cycle. Some slices will have all their data from cardiac systole, other from cardiac diastole. Arteries generally have very slow flow during diastole. For reasons we shall describe shortly, the lumen of these vessels will usually have a higher signal intensity on images obtained during diastole compared to those images obtained during systole. (The higher flow velocity during systole causes loss of signal.)

This same phenomenon can also be observed without intentional EKG-gating, if the TR just happens to be equal to the heart rate (or integer multiples of the heart rate) of the patient. This coincidence of the TR period and the heart rate is called **pseudo-gating**.

If we consider only two-dimensional reconstruction methods, it is possible to identify four distinctly different mechanisms that affect the signal intensity of flowing protons. Two of these mechanisms, **flow-related enhancement** (5, 8) and **even-echo rephasing** (5, 55), produce a high signal intensity within the blood vessel, sometimes referred to as **paradoxical enhancement**. The other two, **time-of-flight** (5, 8, 9) and **flow dephasing** effects (54), produce a "**flow void**", i.e., a loss of signal and therefore a dark appearance.

Flow-related enhancement is an intraluminal high signal intensity occurring when a column of blood (or CSF or other flowing fluid) within the slice being imaged is replenished by "fresh," fully magnetized fluid prior to the start of each new pulse sequence. Fully magnetized fluid has not yet experienced an RF pulse (or at least has not experienced an RF pulse for a period equal to five times T1). The longitudinal magnetization in this fluid will be at its maximum value — at thermal equilibrium.

To produce maximum flow-related enhancement, during the TR interval the fluid must move a distance equal to the length of the vessel within the slice volume. This distance is a minimum (simply the slice thickness) when the flow is perpendicular to the plane of the slice. Thus, flow-related enhancement is best seen when the slice is oriented perpendicular to the flow direction.

Reducing TR will also increase the amount of enhancement. When TR is short, the longitudinal magnetization in the extraluminal, stationary tissue does not have time to fully relax before the start of each new pulse sequence (the tissue is then said to be partially saturated). The MR signal intensity from this tissue is therefore reduced. Fully magnetized blood surrounded by partially saturated tissue will appear very bright against a darker background. Flow-related enhancement is thus most apparent on T1-weighted images.

For flow-related enhancement to occur when using the spin-echo pulse sequence, the fully magnetized blood entering the slice must experience both the 90° and the 180° pulse. If the blood flow velocity is high enough that the column of blood excited by the 90° pulse has left the slice when the 180° pulse is applied to the same slice, no echo will be formed and the vessel lumen will appear dark. This is the time-of-flight effect. We previously said that flow perpendicular to the slice will produce increased MR signal (flow-related enhancement); this is not always true. If a 180° echo pulse is used, vessels oriented perpendicular to the slice will be the most susceptible to the time-of-flight effect; the blood in such vessels will have the shortest distance to travel to escape the slice volume before the application of the 180° pulse. The time-of-flight effect is increased as the flow velocity increases and as TE increases. With a sufficiently large flow velocity or a sufficiently long TE, flow through the slice will yield decreased signal due to the time-of-flight effect.

The amount of signal loss due to the time-of-flight effect is determined strictly by the amount of fluid that has left the slice volume between the 90° and 180° pulses. The fluid that leaves the slice between the 180° pulse and the echo still contributes to the echo signal for the slice even though it is no longer in the slice. The 90° and 180° pulses are normally slice selective but detection of the MR signal is not.

When a multislice spin-echo technique is used, flow-related enhancement is usually not seen in every slice. Only the outer slices of the entire multislice set will receive flow from outside the imaged volume. This flow from outside has not been subjected to RF pulses and is thus "fully magnetized."

Flow-related enhancement in these outer slices will typically occur in only one direction — the direction that brings flow from outside the multislice volume. In an axial multislice spin-echo series through the abdomen, flow-related enhancement can thus be seen in the aorta in the most cranial slice and in the inferior vena cava in the most caudal slice (28). The vessel lumen in deeper slices tends to be replenished by blood that is more or less saturated by RF pulses in the surrounding slices, and therefore do not show flow-related enhancement.

If the velocity of flow is high enough to escape the RF pulses in the outermost slices, fully magnetized blood can be provided to deeper slices prior to each new pulse sequence, and flow-related enhancement can then be observed in these slices. In this case, the flow velocity must not be so high that the blood excited in this deeper slice escapes the slice before the application of the 180° pulse, thus creating a time-of-flight effect.

Flow-related enhancement can be seen in the deeper slices even at relatively low flow velocity if the multiple slices are acquired with a large gap between slices. In that case, blood within a deep slice does not need to be replenished from outside the volume of slices, but can be replenished by fully magnetized blood within the gap — provided the blood spends enough time in the gap to fully regain its longitudinal magnetization.

Flow-related enhancement is very pronounced on single slice images (without multislice technique) obtained with a T1-weighted gradient echo pulse sequence having a short TR (200-400 ms) and a flip angle of 45°-90°. Because there is no 180° pulse, and the gradient used for echo formation is not slice selective, time-of-flight effects producing signal loss will not occur. When these images are obtained as consecutive single slices, each slice, no matter what its location, will have a constant wash-in of fully magnetized blood. With the TR interval and the flip angle selected to provide T1-weighting, the tissue surrounding the vessel lumen will be partially saturated just before each RF excitation pulse. This will reduce the MR signal from this tissue and increase the contrast between the flowing blood and the surrounding tissue. Thin slices oriented perpendicular to the vessels will ensure the most complete replenishment by unsaturated protons and thus show the most pronounced flow-related enhancement.

To understand the last two mechanisms affecting the signal intensity of flowing protons — dephasing effects and even-echo rephasing — the reader should first review the sequence of magnetic field gradients used in 2D FT imaging (Fig. 19-8).

Remember that the x- and z-gradient pulses are designed so that their phase shifting effects cancel out at the time of the echo. However, the net phase shift between protons at different locations produced by the dephasing and rephasing gradients will be zero at the time of the echo only if the protons are stationary with respect to the gradient direction. If the protons move (or flow) along a magnetic field gradient, their precessional frequency will change

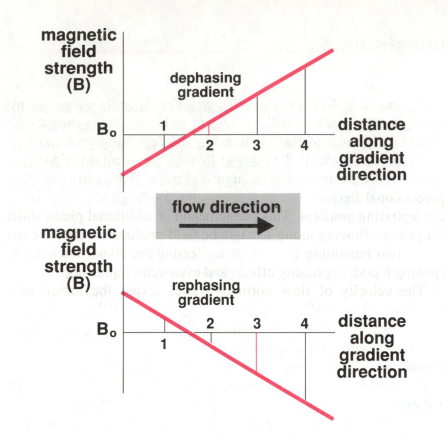

Fig. 22-1. Effect of gradient reversal on stationary and moving protons. Positions 1-4 represent four different locations along the direction of the magnetic field gradient. A dephasing gradient is first applied; in this figure the dephasing gradient is positive. This gradient reduces the magnetic field strength to slightly less than B_0 at position 1, increases it to slightly greater than B_0 at position 2 and further increases it at positions 3 and 4. A rephasing gradient is then applied, in this case as a negative gradient with the same strength and duration as the first gradient. (Here, we are assuming that there is no 180° pulse between these two gradients. If there was a 180° pulse between the gradient pulses, the second gradient would need to be positive to perform its rephasing task. In this case, the reader is encouraged to demonstrate that the results discussed in this figure concerning flow will still occur.) During the negative gradient, the magnetic field strengths relative to B_0 at location 1 through 4 are reversed. The change in phase experienced by a proton during a gradient pulse is proportional to the difference between B_0 and the actual magnetic field at the position of the proton. These differences are indicated by the vertical lines in the figure at each of the four positions.

A stationary proton will experience the same change in phase due to both gradients. However, the phase changes will be in opposite directions and will cancel, yielding no net phase change. If a proton moves from position 2 during the dephasing gradient to position 3 during the rephasing gradient, the change in phase it experiences in those two periods will not be equal (the red vertical lines) and the proton will possess a net phase change after both gradients. The amount of phase change depends on the flow velocity of the proton; a greater velocity produces a greater phase shift. If protons with different velocities exist in the same voxel, they will each experience different phase shifts. Thus they will dephase relative to each other producing a reduced MR signal from that voxel.

as the magnetic field strength changes. For those flowing protons, the rephasing gradient will no longer produce a phase shift that is equal but opposite to that produced by the dephasing gradient (Fig. 22-1). For example, if the protons flow in the direction of the gradient they will experience a larger magnetic field, have a higher precessional frequency, and experience a larger phase shift during the rephasing gradient. This phenomenon of additional phase shifts for protons flowing along a magnetic field gradient forms the basis of the two remaining mechanisms affecting the signal intensity of flowing blood: dephasing effects and even-echo rephasing.

The velocity of flow normally varies across the lumen of a blood vessel or CSF space. Especially in the smaller vessels the velocity profile tends to be parabolic with the peak velocity in the center of the lumen and gradually decreasing velocities towards the vessel wall. Groups of protons that flow along a gradient with different velocities will experience changes in precessional frequency and phase at different rates; consequently they will dephase. The faster the flow or the stronger the gradient, the greater the dephasing. Completely dephased protons cannot induce an MR signal. This dephasing mechanism explains why vessels oriented within the plane of a slice often appear dark; such a flow-void cannot be explained by the time-of-flight effect. When a vessel imaged perpendicular to its long axis appears dark (or has a flow void), the cause can be a combination of dephasing and time-of-flight effects. In CSF, dephasing effects are best seen where the flow channels are narrow and the velocities therefore highest. A patent aqueduct will thus have a signal void due to dephasing even on heavily T2-weighted images, whereas the more stationary CSF is bright.

When a multiecho technique is used, and the echoes are evenly spaced in time, a vessel may show a signal void on the first echo image and a bright intraluminal signal on the second echo image (5, 28, 55). This phenomenon is called even-echo rephasing. Its occurrence is best seen at laminar, linear flow, and it is therefore most commonly observed in veins (55). **Laminar flow** means that the lumen of the vessel can be divided into thin laminae or shells, each lamina having its own flow velocity, different from the other laminae. **Linear flow** implies that the velocity is non-pulsatile and constant. The flow profile is usually parabolic with the highest velocity in a central core, which is surrounded by thin laminae, one outside the other, having ever decreasing velocities towards the periphery of the lumen.

When protons (within one lamina) are moving along a linear magnetic field gradient with a constant velocity (v), their precessional frequency will experience a linear increase with time; the rate of increase will be proportional to their velocity. Their total phase shift ($\Delta\phi$), i.e., the angle of gain or loss in phase, during a time period t, is then determined by the formula:

$$\Delta\phi = \frac{1}{2}\gamma \, Gvt^2 \qquad\qquad (22\text{-}1)$$

where γ is the gyromagnetic ratio and G is the strength of the magnetic field gradient. An understanding of the entire formula or its derivation is not important in this context; what should be appreciated, however, is that the phase shift is proportional to both v and t^2.

For an intuitive understanding of this relationship between $\Delta\phi$, v, and t, let us use a simple example in which protons are moving along the gradient in the direction of increasing field strength. We will look at the protons at three evenly spaced points in time: $t = 0$, $t = t_0$, and $t = 2t_0$. Suppose one group of protons is moving with twice the velocity of a second group of protons. In the time from $t = 0$ to $t = t_0$, the faster protons have traveled twice the distance of the slower protons. Thus the increase in the magnetic field experienced by the faster protons during this time is twice that of the slower protons. This means that the change in precessional frequency is twice as great for the faster protons compared to the slower protons and that the phase shift is also twice as great for the faster protons compared to the slower protons. This confirms the dependence of the phase shift on v.

If we now continue to observe the protons until $t = 2t_0$, we see that the change in magnetic field experienced by both the faster and slower protons has doubled from the time t_0 to the time $2t_0$. The total phase shift experienced by either group of protons in the time from $t = 0$ to $t = 2t_0$ will be *four times* the phase shift each experienced from $t = 0$ to $t = t_0$. This 4X increase in the phase shift is due to the combination of two separate factors, each of which causes a 2X increase in the phase shift: (1) During the $t = 0$ to $t = 2t_0$ period the change in magnetic field strength is double the change experienced during the $t = 0$ to $t = t_0$ period. Thus the change in precessional frequency during the $t = 0$ to $t = 2t_0$ period is double the change experienced during the $t = 0$ to $t = t_0$ period. If we calculate the average precessional frequency during the time intervals 0 to t_0 and 0 to $2t_0$ and subtract the precessional frequency at $t = 0$, this difference for the interval 0 to $2t_0$ will be twice that for the interval 0 to t_0. (2) During the period $t = 0$ to $t = 2t_0$ this doubled average increase in precessional frequency has twice as long to act compared to the $t = 0$ to $t = t_0$ period. The total phase shift during a time interval equals the average increase in precessional frequency during the interval times the length of the time interval. This yields a 4X increase in the total phase shift.

Let us now consider two groups of protons, A and B, from different laminae within a vessel having linear, laminar flow. Like our previous example, flow is directed parallel to a magnetic field gradient. In this case, to make our computations easy, we assume that the flow is along the x direction and that group B has twice the velocity of group A. According to Eq. 22-1, the group B protons will have twice the phase shift of the group A protons during any time interval when the gradient is applied (i.e., $\Delta\phi_B = 2\Delta\phi_A$) If a multiecho spin-echo sequence with equally spaced echoes is applied, the time from the 90° pulse to the second echo can be divided into four equally long time intervals (Fig. 22-2). We will call the time duration of each interval τ.

Phase changes will only occur while the x-gradient is applied. The x-gradient is not on all the time; however, because of the nearly symmetrical placements of the gradient pulses we can

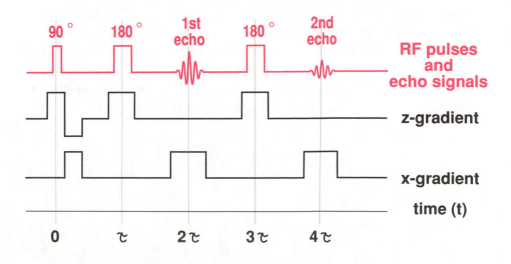
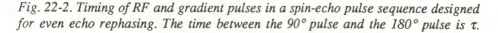

Fig. 22-2. Timing of RF and gradient pulses in a spin-echo pulse sequence designed for even echo rephasing. The time between the 90° pulse and the 180° pulse is τ.

assume in our example that the gradient is on all the time, without invalidating the conclusions. If the x-gradient is positive and the flow is in the positive x direction, the phase shift of the flowing protons will always increase, at an increasing rate, as time elapses (compared to protons remaining stationary) (Eq. 22-1). However, each 180° pulse will invert the phase shift that has accumulated since the 90° pulse.

If we ignore the effect of the 180° pulse for the moment, we can calculate the phase shift that occurs at the end of each of the 4 specified time periods. Let's designate the total phase shift during the first time interval, i.e., from the 90° pulse to the first 180° pulse by the symbol a. From Eq. 22-1 we use the fact that the phase shift is proprtional to t^2. At the first 180° pulse, a time τ has passed since the 90° pulse. At the first echo, a time 2τ has elapsed since the 90° pulse. The total phase shift from the 90° pulse to the first echo is thus $2^2a = 4a$. Similarly, after an elapsed time of 3τ (at the second 180° pulse), the total phase shift will be $3^2a = 9a$, and after an elapsed time of 4τ (at the second echo), the total phase shift since the 90° pulse will be $4^2a = 16a$. We now can look at what happens to the phase shifts of the group A and B protons during the four time intervals. Again, just to make it easy for ourselves, we will assume that the phase gain of the group A protons is 10° during the first time interval. The phase gain of the group B protons (with twice the velocity of group A) is consequently 20° during the same period. At the end of the second interval ($t = 2\tau$) the group A protons have a total phase gain of 40°, thus 30° of phase shift occurred during second interval. At $t = 2\tau$ the group B protons have a total phase gain of 80°, thus 60° of phase shift occurred during the second interval. The total phase shifts of the two groups at the end of each of the four intervals and the amount of phase shift occurring within each interval is summarized in Table 22-1.

Table 22-1.

	$t = 0$	$t = \tau$	$t = 2\tau$	$t = 3\tau$	$t = 4\tau$
Total Phase Shift					
Group A	0°	10°	40°	90°	160°
Group B	0°	20°	80°	180°	320°

Phase Shift Occurring Within the Interval Indicated	0 to τ	τ to 2τ	2τ to 3τ	3τ to 4τ
Group A	10°	30°	50°	70°
Group B	20°	60°	100°	140°

So far we have ignored the phase inverting effect of the 180° pulse. If we now include the effects of the 180° pulse, the resulting phase shifts at the end of each interval are shown in Table 22-2. When the 180° pulse is applied, the phase will invert or change sign (i.e., change from leading to lagging). After the 180° pulse the phase continues to increase in the same fashion as before: the increase in phase during each interval remains the same as in Table 22-1.

Table 22-2.

	$t = 0$	$t = \tau$	$t = 2\tau$	$t = 3\tau$	$t = 4\tau$
Total Phase Shift					
Group A	0°	10/-10°	20°	70/-70°	0°
Group B	0°	20/-20°	40°	140/-140°	0°

In Table 22-2 a pair of numbers are given at $t = \tau$ and $t = 3\tau$, the times of the 180° pulses. The first number is the total phase shift just before the 180° pulse, the second number is the total phase shift just after the 180° pulse. To demonstrate how we obtained the numbers in the table, we shall show the procedure for the group A protons. At $t = \tau$, just before the 180° pulse the total phase shift = 10°; just after the 180° pulse the total phase shift is -10°. During the second interval the phase shift has increased by 30° (from Table 22-1), yielding a total phase shift of 20° (-10°+ 30° = 20°) at $t = 2\tau$.

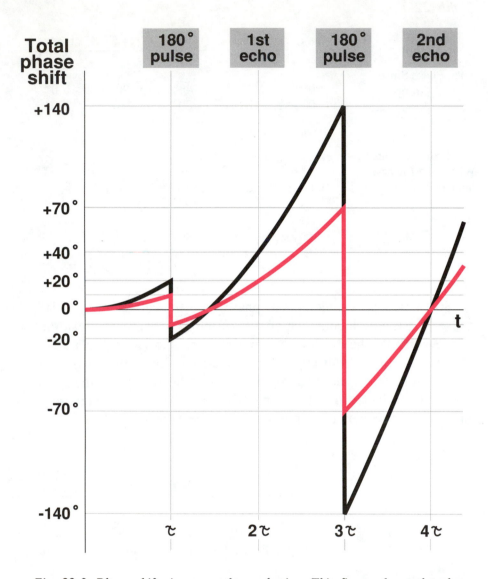

Fig. 22-3. Phase shifts in even-echo rephasing. This figure shows the phase shifts at different times for protons flowing at 2 different velocities. The pulse sequence of Fig. 22-2 is used. The red curve represents the slower flowing protons, group A in the text. The black curve represents the faster flowing protons, group B in the text. Group B protons are flowing at twice the speed of group A protons. Here we will assume that the speed of the group A protons is such that a 10° phase shift is produced in the time τ. The data from this curve at times 0, τ, 2τ, 3τ, and 4τ are compiled in Table 22-2. At the first echo the protons with different flow velocities have different phase shifts. This relative dephasing will reduce the strength of the first echo signal. At the second echo, the phase difference between protons with different flow velocities has been eliminated. The strength of the second echo will thus be larger than that of the first echo.

During the third interval the phase shift increases by 50° (from Table 22-1) yielding a total phase shift of 70° just before the 180° pulse; just after the 180° pulse the total phase shift is -70°. During the third interval the phase shift increases by 70° (from Table 22-1) yielding a total phase shift of 0 at t = 4τ.

Fig. 22-3 is a graphical representation of the numbers given in Table 22-2. This figure plots the total phase shift as a function of time. In Table 22-2 and Fig. 22-3 we see that at t = 2τ the phase shift of group A protons is different than the phase shift of group B protons (20° versus 40°). Thus the flowing protons are dephased at the first echo (**odd-echo dephasing**). At t = 4τ the phase shift of group A protons is the same as the phase shift of group B protons, thus the flowing protons are in phase at the second echo (even-echo rephasing). If we continued this process for further successive echoes we would find that the protons would be dephased for the odd-numbered echoes and would be in phase for the even-numbered echoes as long as the time intervals between the echoes are equal.* In our example, we would have obtained the same basic result if we had chosen different phase shifts due to different flow velocities or different gradient strengths.

Even-echo rephasing is best seen with slow flow. When the velocity of flow is high, the intensity of the even-echo will be reduced due to time-of-flight effects, particularly if the vessel is oriented perpendicular to the slice being imaged. (We have already mentioned that even-echo rephasing is enhanced with uniform flow as opposed to pulsatile flow.) The observation of even-echo rephasing may be of help in differentiating between slow flow and thrombosis. In some cases, slow flow and thrombus may appear equally bright in the first echo image. If a second, even-echo is obtained, the thrombus will appear less bright than on the first echo image due to T2 relaxation in the stationary tissue. Slow flow, however, will appear even brighter on the second echo image due to even-echo rephasing.

---

* Of course, the even-numbered echoes still reflect proton dephasing due to T2 processes. In addition, non-uniformities in the static magnetic field and in the applied gradient field will produce additional dephasing. (Remember that the 180° pulse only compensates for magnetic field inhomogeneities when the protons are stationary and thus experience the same magnetic field before and after the 180° pulse.) Though the signal from the even-echoes will be stronger than those from the odd-echoes, the amplitude of the each successive even-echo will still decrease in time due to the above dephasing effects.

We have seen how fluid motion can interfere with the desired dephasing and rephasing effects of the applied magnetic field gradients. It is possible, however, to design the strength, duration and timing of the gradients so that the additional, unwanted dephasing caused by constant velocity or even accelerated flow is compensated. The use of these specially designed gradients is called **gradient moment nulling** or **flow compensation** (49, 56, 59, 60). A flow compensated spin-echo image will not have the "usual" flow voids. Areas of blood, CSF, and other fluid flow that would normally appear dark due to dephasing effects will image brighter with flow compensation. Flow compensation techniques can be used with both conventional and gradient echo imaging.

Flow can interfere with image quality outside of a vessel by causing **image artifacts** (22, 26, 34). We have seen how flow can induce phase shifts in the moving protons. In Chapter 19 we saw how phase shift is used to encode locations in the y direction. Errors in phase due to fluid motion can, therefore, produce **ghost images** of vessels. These ghost images are displaced in the phase-encoding (y) direction from the true vessel position. Flow artifacts can be particularly disturbing when the vessel lumen signal is strong due to flow-related enhancement; such conditions are frequently found when using fast scan gradient echo imaging. Flow compensation can significantly reduce these flow artifacts.

Flow compensation can reduce the contrast between the vessel lumen and surrounding tissue by eliminating the signal void due to dephasing. However, if the same image is obtained twice with the same parameters, first with flow compensation, then without flow compensation, a subtraction of the two images will show only the vessels (49). This is one of several methods that can be used to make **MR angiograms**.

23. Effects Of Chemical Shift In MRI

In Chapter 9 we saw that even if the MR magnet produces a perfectly uniform magnetic field, the protons in the tissue will not experience a constant magnetic field equal to the uniform field of the MR magnet. In addition to the static magnetic field of the MR magnet they will experience a much smaller varying magnetic field due to the magnetic moments of the protons and molecules in the tissues; we called this magnetic "noise."

In Chapter 9 we also saw that the protons in the body are subject to electronic shielding. The static magnetic field induces an electric current in the "clouds" of electrons surrounding the protons (and other nuclei). These currents produce a weak magnetic field that opposes the static magnetic field. This reduces the local magnetic field experienced by the nucleus within the "cloud." While the strength of this shielding can vary somewhat as the molecule tumbles with different orientations, there will still be an average reduction in the magnetic field at the nucleus. This reduction in the local magnetic field experienced by the nucleus causes the phenomenon called **chemical shift**.

The amount of shielding experienced by the protons will depend upon the molecule or atom to which the proton is attached. When the proton is attached to an electronegative atom like oxygen, the shielding is relatively weak, because oxygen draws the electron cloud away from the proton. If the proton is attached to carbon, the shielding is stronger because carbon is less electronegative then oxygen. The stronger the shielding, the weaker the field strength experienced by the proton and the slower its precessional frequency. Protons in fat molecules (carbon-attached) consequently have a slower precessional frequency than protons in water molecules (oxygen-attached). This difference in precessional frequency according to molecular attachment is called chemical shift. The actual frequency difference measured in hertz is proportional to the strength of the static magnetic field. At 0.35 T protons will precess with a frequency of about 15 MHz. However, the protons in water molecules will have a precessional frequency that is about 50 Hz higher than that of the **methylene protons** ($-CH_2-$) in fat. At 1.5 T, when the precessional frequency of protons is approximately 64 MHz, the difference in precessional frequency between water and methylene protons increases to approximately 210 Hz. The

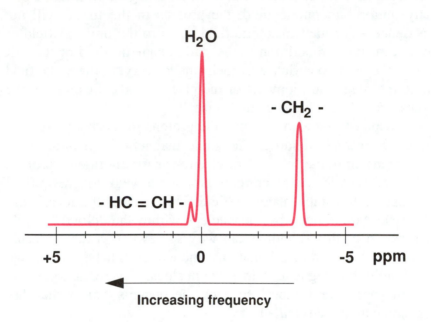

Fig. 23-1. MR spectrum and chemical shift. This figure gives a schematic representation of the MR spectrum of hydrogen, 1H, for tissue containing both water and fat. The spectrum is a plot of the strength of the MR signal at different frequencies in a constant magnetic field. By convention, increasing frequency is plotted to the left. Each peak in the spectrum is produced by protons in distinct chemical environments. Protons with differences in chemical bonding or in position relative to the rest of the molecule will experience slightly different local magnetic fields due to the effects of electronic shielding. These protons therefore precess (and produce MR signals) at slightly different frequencies. The difference in frequency between 2 different chemical forms is called their chemical shift. This frequency difference is commonly expressed in parts per million (ppm) of the precession frequency. Using this measure, the chemical shift does not depend on the magnetic field strength.

The zero point of chemical shift is arbitrary, though standards are often chosen so that the chemical shifts of other chemical forms are given relative to the standard. In the graph above, the principle components of the spectrum are the protons in water and the methylene protons (-CH₂-) in fat. The chemical shift between these spectral components is about 3.3 ppm. A minor component of fat, the vinyl (-HC=CH-) protons in unsaturated fatty acids, is also shown in this spectrum. It has a chemical shift of about 3.7 ppm relative to the methylene protons.

chemical shift is more commonly expressed in **parts per million (ppm)** of the precessional frequency, since this unit is independent of field strength. The chemical shift in the above example is 50 Hz per 15 million Hz or 210 Hz per 64 million Hz, both of which are 3.3 ppm.

Chemical shift forms the basis of **MR spectroscopy**. In spectroscopy we are interested in determining the types and amounts of the chemical forms of a particular nucleus (such as hydrogen, sodium, or phosphorus) present in a sample. This can be

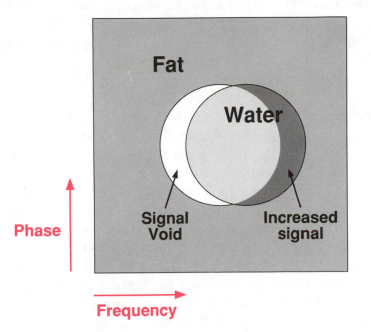

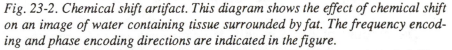

Fig. 23-2. Chemical shift artifact. This diagram shows the effect of chemical shift on an image of water containing tissue surrounded by fat. The frequency encoding and phase encoding directions are indicated in the figure.

In MR imaging, position in the readout direction is frequency encoded. Since fat and water protons precess at slightly different frequencies, the MR signals of fat and water protons in the same voxel will contribute to different positions in the image; this shift in position will be in the readout direction. When in the same static magnetic field, water protons precess at a higher frequency than fat protons. If the readout gradient causes the magnetic field to increase to the right, the image reconstruction program will interpret the higher frequency of the water protons as a shift in position to the right, relative to fat tissue. This shift will produce an increased signal at water-fat boundaries where fat is to the right of water, since the fat and water signals from voxels on either side of the boundary will be assigned to the same pixels in the image. At boundaries with fat to the left of water the water and fat signals will be displaced apart in the image, leaving pixels with signal contributions from neither; this creates a signal void at the boundary.

accomplished by exciting the nuclei in the sample and then analyzing the resulting MR signal (using 1D FT) to determine the frequencies it contains. Since the different chemical forms of the nucleus will resonate at slightly different frequencies, the MR spectrum contains a chemical analysis of the sample for that particular element. We have just seen that the protons in water and in methylene groups resonate at slightly different frequencies. Since these are the principle chemical forms in which hydrogen is found in the body, the MR frequency spectrum obtained from human tissue contains two major peaks: one from the water protons, the other from the methylene protons that are the major spectral component of fat. The distance between these two peaks in the spectrum is approximately 3.3 ppm; this is their chemical shift. An example of a proton spectrum of tissue is given schematically in Fig. 23-1.

In conventional MR imaging the protons from water and fat both contribute to the image since both are excited by the initial RF pulse of a pulse sequence. The contribution of both water and fat protons to the MR signal gives rise to the **chemical shift artifact** (Fig. 23-2). Since frequency encoding is used to determine the x position in MRI, the change in frequency caused by chemical shift will be converted into a shift in position along the readout (x) direction. If the readout gradient points in the positive x direction, then a voxel containing higher frequency water protons will be shifted to the right in the image relative to a voxel containing fat protons. The computer has no way of knowing that the higher frequency of the water signal is due to chemical shift and not due to a shift in position along the x direction. Thus the higher frequency from the water voxel will be interpreted as a larger x-coordinate by the computer.

The amount of position shift is determined by the strengths of the readout gradient and the static magnetic field. Increasing the gradient strength or decreasing the static magnetic field strength will reduce the amount of spatial misregistration produced by the chemical shift artifact. Suppose the gradient strength is 1 mT/m in the positive x direction and the static magnetic field strength is 1.5 T. With a chemical shift of 3.3 ppm, the difference in magnetic field felt by the water and fat protons is $(3.3 \times 10^{-6})(1.5\ T) = 5 \times 10^{-6}\ T$ or 0.005 mT. This corresponds to a shift in position of 5 mm when using a 1 mT/m (0.001 mT/mm) readout gradient. Thus if a pure water voxel and a pure fat voxel were placed next to each other in the patient, such that both voxels have the same x-coordinate, in the image the water voxel would appear 5 mm to the right of the fat voxel. If the static field strength were reduced by 1/3 to 0.5 T, the difference in the magnetic field felt by the water and fat protons would also decrease by 1/3; this in turn would decrease the position shift by 1/3 to 1.7 mm. Increasing the gradient strength by a factor of 3 to 0.003 mT/mm would also decrease the position shift to 1.7 mm.

Due to the relative spatial shift of the images of fat and water containing tissues, the chemical shift artifact is apparent in MR images as a bright or dark band at the boundary between tissues composed mainly of water and those composed mainly of fat. In the above example since water is shifted to the right compared to fat, a boundary with water containing tissue on the left and fat on the right will appear bright in the image. This is due to the super-position of the water and fat signals in the pixels at the boundary.

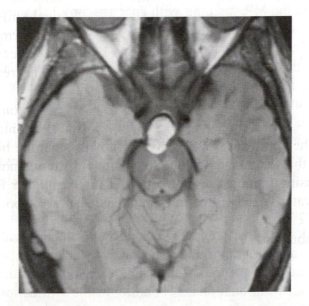

Fig. 23-3. Clinical image showing the chemical shift artifact. This axial proton density weighted image (TR = 2,000 ms, TE = 20 ms) demonstrates a fatty tumor (supracellar cystic craniopharyngioma) with a high signal intensity due its content of fat. The signal from the surrounding CSF space and brain tissue is derived almost exclusively from water protons. The readout gradient is oriented in the occipito-frontal direction, the vertical direction in this image. The chemical shift artifact is consequently seen as a black band frontal to the fat containing tumor (above it) and as a bright band on the opposite side of the tumor (below it).

A boundary with water containing tissue on the right and fat on the left will appear dark in the image. Since the edge of water signal is moved to the right, away from the edge of the fat signal there will remain some area in the image between the water and fat signals which will have contributions from neither. A clinical example of chemical shift artifact is shown in Fig. 23-3.

Just as a large magnetic field gradient is helpful in reducing the spatial shift produced by the chemical shift in the readout direction, a large slice select gradient helps to reduce the effect of chemical shift on slice selection. When an excitation RF pulse is applied during a slice selection gradient, both water and fat protons are excited within a slice, however, the position of the slice is not exactly the same for the water and fat protons. The chemical shift between water and fat produces a shift in the location of the slice position in the same way it introduces a shift in position in the readout direction.

Because of the frequency encoding in the x direction, it is difficult to combine MR imaging with spectroscopy using the imaging techniques we have previously discussed. However, if we do away with the frequency encoding, frequency in the MR signal can directly indicate chemical shift instead of position. This can be done in 2D imaging, without losing the ability to locate signals in both the x and y direction, by using the following technique.

The slice selection z-gradient and phase encoding y-gradient are applied in the normal fashion. However, a phase encoding gradient is also used in the x direction, as a replacement for the readout gradient. (We have already seen how two separate dimensions can both be phase encoded: this was done for the y and z directions in our explanation of the 3D imaging method.) This technique pays a great penalty in time, however. For normal imaging with a 256 x 128 image matrix, a minimum of 128 MR signals must be obtained using 128 pulse sequences, due to phase encoding in the y direction. With an additional phase encoding in the x direction, each of these 128 pulse sequences must be repeated 256 times with varying x-gradients to accomplish the phase encoding in the x direction. This increases the imaging time by a factor of 256. Obviously this technique is not practical for normal clinical scanning, and is only usable with a greatly reduced image matrix size (eg., 64 x 64) and thus a reduced resolution. What it does provide, however, is an MR spectrum for each pixel in the image.

It is quite possible to make use of chemical shift effects in imaging without obtaining complete spectral information for each point in the image. For example, it is possible to image only those nuclei having a particular chemical shift (such as methylene protons or water protons), or to produce images whose image densities are affected by the chemical shifts of the nuclei. Such methods are called **chemical shift imaging**. Chemical shift imaging of protons is simplified due to the presence of only two major spectral components in most tissues: methylene protons in fat and water protons. Various techniques have been developed that can capture the signals from only the water or only the fat protons (2, 10). Such methods are often called **fat suppression** and **water suppression** techniques respectively.

In Chapter 15 we described a pulse sequence that can be used for fat suppression even though it does not actually make use of the chemical shift effect. The STIR (short TI inversion-recovery) pulse sequence can suppress the MR signal from fat if TI (the interval between the 180° and 90° pulses) is selected to be approximately equal to 0.69 T1 for fat. After the longitudinal magnetization is inverted by the 180° pulse, T1 relaxation causes the magnetization to decrease to zero from its maximum negative value and to then become positive and increase toward its equilibrium value, M_0 (Fig. 15-2). At a time 0.69 T1 after the 180° inversion pulse the magnetization is equal to zero. With STIR, TI must be selected so that for all tissues 0.69 T1 is equal to or greater than TI; this insures that no tissues have positive longitudinal magnetizations at the time of the 90° pulse. Since the T1 of fat is substantially shorter than that of most other tissues, the requirements of the STIR method can be met while suppressing the fat signal.

One class of chemical shift imaging techniques uses selective excitation of only the water or fat protons. With these techniques a spin-echo pulse sequence is often used in which the 90° or 180° pulse is made non-selective for slice selection (the slice selection gradient remains off during the RF pulse). The bandwidth of this pulse is then made quite narrow so that it affects only the protons with the desired chemical shift. One of the RF pulses accomplishes the slice selection while the other selects the proton type (e.g., water or fat) to be imaged.

Other chemical shift imaging techniques use **selective saturation**. One method is to apply a narrow bandwidth 90° pulse, selectively saturating the nuclei whose signals are not wanted. This 90°

pulse destroys the longitudinal magnetization of these protons; a dephasing gradient pulse is then applied to destroy their transverse magnetization. At this point a normal pulse sequence can be applied and only the desired nuclei will be imaged. In the above techniques, the use of narrow bandwidth RF pulses to excite or saturate a particular spectral component requires a very homogeneous magnetic field to assure that the Larmor frequency of the nucleus in question does not vary too much due to magnetic field inhomogeneities.

One relatively simple method of chemical shift imaging which does not require such high homogeneity of the static magnetic field will be described in a bit more detail. In this method both water and fat protons are affected by the RF pulses and contribute to the MR image.

In normal spin-echo imaging we saw how the chemical shift produces a misregistration artifact.Chemical shift effects can also produce phase shifts that can affect the signal intensity within voxels containing both water and fat. After the 90° pulse, the water and fat protons located at the same position in the patient will be precessing at different frequencies due to chemical shift. Therefore they will experience a relative phase shift. The effect of this phase shift is not normally noticed in the echo signal due to the rephasing effects of the 180° pulse. Just as explained for protons precessing with constant but different frequencies due to field inhomogeneities caused by the MR magnet (Fig. 13-1), protons precessing with constant but different frequencies due to chemical shift will dephase prior to the 180° pulse and rephase after the pulse so that they are back in phase at the spin echo. This assumes that the standard spin-echo pulse sequence is used, in which the gradient echo coincides with the Hahn echo produced by the 180° pulse. The image obtained in this way is called the **in-phase image** since the magnetization due to the fat and water are in phase within each voxel. If, however, we change the timing of the 180° pulse so that the gradient echo and the Hahn echo occur at different times, optimum rephasing of the chemical shift dephasing will not occur at the time of the gradient echo.

With a static magnetic field of 1.5 T, the difference in precessional frequency between water and fat is about 210 Hz. If the precession of the protons is observed in a frame of reference rotating at the Larmor frequency of the fat protons, the magnetic moments of the fat protons will appear stationary in this frame and the magnetic moments of the water protons will rotate with a frequency of about 210 Hz in the clockwise direction as viewed from above the z-axis. Each rotation of the transverse component of the magnetic moment of a water proton will take 1/210 sec or 4.8 ms. In half that time (2.4 ms) the proton will have rotated half way around. If the Hahn echo occurs 2.4 ms before (or after) the gradient echo, the magnetization due to the protons from fat and water will be 180° out of phase at the time of the gradient echo, i.e., they will point in opposite directions. The image obtained in this case is called the **opposed image** or **out-of-phase image**.

The in-phase and opposed images will have differences in image contrast. In voxels containing *both* water and fat, the signal intensity will be normal in the in-phase image, since the transverse magnetization due to the water and fat point in the same direction and will add to give a larger net transverse magnetization. The signal intensity will be decreased in the opposed image since, in this case, the transverse magnetization due to the water and fat will tend to cancel. The voxels at a boundary between fat and water containing tissue will contain both fat and water protons. This means that the boundary between fat and muscle will be marked by decreased density in the opposed image.

In voxels containing *only* water or fat, the signal intensity should not be significantly different with in-phase and opposed images. In these two images the transverse magnetization in corresponding voxels differs only in its phase, not in its size. The brightness of a pixel depends only on the strength of the MR signal from its corresponding voxel, not on its phase. The strength and phase of the MR signal is derived directly from the size and phase of the transverse magnetization. Thus the pixel brightness is not affected by the phase of the voxel magnetization.

We can use the in-phase and opposed images to form a **water image** containing signals from only the water protons, and a **fat image** containing signals from only the fat protons (10, 12, 37). With the in-phase image, the transverse magnetization in each voxel is the in-phase sum (w + f) of the transverse magnetization of the water protons (w) and that of the fat protons (f). In the opposed image, the transverse magnetization in each voxel is the out-of-phase sum (w - f) of the transverse magnetization of the water protons and that of the fat protons. A water image will be the sum of the in-phase and opposed images: (w + f) + (w - f) = 2w. A fat image is obtained by subtracting one from the other: (w + f) - (w - f) = 2f. Such chemical shift images have been found of special value in the imaging of fatty infiltration of the liver (37). The "opposed" image by itself may also offer increased detectability of liver metastases (38).

The above techniques of producing in-phase and opposed images can also be applied to gradient echo imaging. Since these techniques do not use a 180° RF pulse, dephasing of the water and fat protons proceeds unimpeded from the time of the initial RF excitation pulse. According to our previous calculations, if the static magnetic field strength is 1.5 T, the magnetization due to water and fat theoretically should be in phase at 4.8, 9.6, 14.4, and 19.2 ms after the initial RF pulse. The magnetization should be out of phase at 2.4, 7.2, 12.0, and 16.8 ms after the initial RF pulse. (Experimentally we may find that these values need a slight adjustment: a small constant may need to be added to or subtracted from each time. However, the differences between successive in-phase and out-of-phase conditions will remain at 2.4 ms.) If the read-out gradients are adjusted so that images are obtained with a TE of 9.6 and 12.0 ms, for example, the first image will be an in-phase image and the second image an opposed image. These images will have much the same properties and use as those just described using the spin-echo pulse sequence. They can be combined to form water and fat images.

Another type of in-phase and out-of-phase effect may occur when imaging with fast scan techniques using gradient echoes (56, 57, 61). In the references just cited it has been noted that very small increments of TE (about 2.1 ms at 1.5 T) can markedly change the signal intensity of adipose tissue. The explanation given for this variation involves the different chemical environments experienced by the protons in the long-chain fatty acids of adipose tissue. The major spectral component of fat is the methylene ($-CH_2-$) protons, but a minor component present in unsaturated fatty acids, the **vinyl** ($-HC = CH-$) protons can also contribute to the MR signal in noticeable amounts (57, 61). There is a chemical shift of 3.7 ppm between these two components, with the vinyl protons having a slightly higher precessional frequency than water protons (56, 57, 61). This yields a difference in precessional frequency of 240 Hz at 1.5 T.

Using reasoning identical to that used for water vs. fat, we find that the magnetization due to these two components of fat should be in phase every 1/240 sec or 4.2 ms. Thus the in-phase condition will theoretically occur at 4.2, 8.4, 12.6, and 16.8 ms after the initial RF pulse. The out-of-phase condition should occur at 6.3,

10.5, 14.7, and 18.9 ms after the initial RF pulse. (As before, we may find experimentally that these values need a slight adjustment: a small constant may need to be added to or subtracted from each time. However, the differences between successive in-phase and out-of-phase conditions will remain at 2.1 ms.) If the read-out gradients are adjusted so that images are obtained with a TE of 12.6 and 14.7 ms, for example, the first image will be an in-phase image and the second image an opposed image (for the components of fat).

In the opposed image, a decrease in the brightness of adipose tissue is observed, compared to the in-phase image. This decrease is due to the partial cancellation of the transverse magnetization from the methylene protons by the out-of-phase magnetization of the vinyl protons. In the opposed image a dark boundary is sometimes seen between muscle and fat (57, 61). This is due not to the out-of-phase relationship in the image between methylene and vinyl protons, but due to a coincidental nearly out-of-phase relationship between water and methylene protons. Because the phase shifts between water and methylene and between methylene and vinyl are similar (3.3 ppm versus 3.7 ppm) it is possible to produce images that are nearly in-phase or out-of-phase for both water/methylene and methylene/vinyl if the echo times are selected carefully.

Symbols and Abbreviations

α — The flip angle: the angle of the RF excitation pulse.

γ — The gyromagnetic ratio (also called the magnetogyric ratio).

θ — The angle between the magnetization vector and the static magnetic field vector (along the z-axis).

ϕ — The phase angle of the magnetization vector. The phase angle of a sine wave signal.

μ — Magnetic moment vector.

ω — Angular frequency (radians/second).

ω_0 — The resonance frequency or Larmor frequency expressed as an angular frequency (radians/second).

ω_1 — The angular frequency of precession of the magnetic moment of magnetization around the B_1 magnetic field in the rotating frame of reference.

τ — The time from the 90° pulse to the 180° pulse in a spin-echo pulse sequence.

A — The amplitude of a sine wave signal.

B — Magnetic field.

B_0 — The static magnetic field produced by the MR magnet.

B_1 — The component of the radio-frequency (RF) oscillating magnetic field that rotates in the same direction as the precession of the magnetization.

e — The base of the natural logarithm; equal to approximately 2.7.

E — Energy. The energy of a photon.

ΔE — The energy difference between parallel and antiparallel protons.

E-M — Electromagnetic.

f — Frequency (Hertz = cycles/second).

f_0 — The resonance frequency or Larmor frequency expressed in Hertz.

FID — The free induction decay: the MR signal after the application of a 90° pulse.

FT — Fourier transform.

G — The strength of the magnetic field gradient (Tesla/meter).

h — Planck's constant.

$\hbar$ — h/2*

I — Spin

IR — Inversion-recovery pulse sequence.

k — Boltzmann constant.

M — The magnetization.

M_0 — The magnetization at thermal equilibrium.

M'_0 — The magnetization present at the start of a pulse sequence.

M_{xy} — The transverse magnetization: the component of the magnetization in the x-y plane.

M_{xy}^H — The transverse magnetization that would exist if the static magnetic field were perfectly homogeneous.

M_z — The longitudinal magnetization: the component of the magnetization along the z-axis (the direction of the static magnetic field).

MR — Magnetic resonance.

MRI — magnetic resonance imaging.

N_p — Proton density: the number of protons per unit mass (protons/kg).

$N_{antiparallel}$ — The number of antiparallel protons per unit mass.

$N_{parallel}$ — The number of parallel protons per unit mass.

NEX — Number of excitations: number of pulse sequence performed for each "view" in the production of a single MR image.

ppm — Parts per million: unit of chemical shift.

PS — Partial-saturation pulse sequence.

SE — Spin-echo pulse sequence.

SR — Saturation-recovery pulse sequence.

STIR — Short T1 inversion-recovery pulse sequence.

τ — Time

T — Absolute temperature (K*).

TD — Delay time: the time between the last echo of a pulse sequence and the first RF pulse of the next pulse sequence at the same selected slice or volume.

TE — Echo time: the time between the 90° pulse and the mid-point of the echo in a spin-echo or inversion-recovery pulse sequence.

TI — Inversion time: the time between the 180° inversion pulse and the 90° pulse in an inversion-recovery pulse sequence.

TR — Repetition time: the time between the beginning of a pulse sequence and the beginning of the next pulse sequence at the same selected slice or volume.

T1 — Longitudinal relaxation time; thermal relaxation time; spin-lattice relaxation time.

T2 — Transverse relaxation time; spin-spin relaxation time.

T2* — The time constant of the FID signal: the time over which the amplitude of the FID decreases to 37% of its initial maximum value.

x, y, z — The coordinate axes of the stationary coordinate system.

x', y', z — The coordinate axes of the rotating coordinate system.

1D FT — One-dimensional Fourier transform.

2D FT — Two-dimensional Fourier transform.

3D FT — Three dimensional Fourier transform.

1. Abragam A: *Principles of Nuclear Magnetism*. Oxford University Press, Oxford, 1961.

2. Axel L: Chemical shift imaging. In: Stark DD, Bradley WG   (eds). *Magnetic Resonance Imaging*: 201-228. C.V. Mosby, St. Louis, 1988.

3. Bloch F: Nuclear induction. *Phys Rev* 70: 460-474, 1946.

4. Bradley WG: Pathophysiologic correlates of signal alterations. In: Brant-Zawadzki M, Norman D (eds). *Magnetic Resonance Imaging of the Central Nervous System*: 23-42. Raven Press, New York, 1987.

5. Bradley WG: Flow phenomena. In: Stark DD, Bradley WG   (eds). *Magnetic Resonance Imaging*: 108-137. C.V. Mosby, St. Louis, 1988.

6. Bradley WG: MRI of hemorrhage and iron in the brain. In: Stark DD, Bradley WG (eds). *Magnetic Resonance Imaging*: 359-374. C.V. Mosby, St. Louis, 1988.

7. Bradley WG, Schmidt PG: Effect of methemoglobin formation on the MR appearance of subarachnoid hemorrhage. *Radiology* 156: 99-103, 1985.

8. Bradley WG, Waluch V: Blood flow: Magnetic resonance imaging. *Radiology* 154: 443-450, 1985.

9. Bradley WG, Waluch V, Lai K, Fernandez E, Spalter C: The appearance of rapidly flowing blood on magnetic resonance images. *AJR* 143:1167-1174, 1984.

10. Brateman L: Chemical shift imaging: A review. *AJR* 146: 971-980, 1986.

11. Bydder GM, Young IR: MR imaging: Clinical use of the inversion recovery sequence. *J Comput Assist Tomogr* 9: 659-675, 1985.

12. Dixon WT: Simple proton spectroscopic imaging. *Radiology* 153: 189-194, 1984.

13. Engelstad BL, Wolf GL: Contrast agents. In: Stark DD, Bradley WG (eds). *Magnetic Resonance Imaging*: 161-181. C.V. Mosby, St. Louis, 1988.

14. Farrar TC: *An Introduction to Pulse NMR Spectroscopy*. Farragut Press, Madison, 1987.

15. Frahm J, Haase A, Matthaei D: Rapid three-dimensional NMR imaging using the FLASH technique. *J Comput Assist Tomogr* 10: 363-368, 1986.

16. Fukushima E, Roeder SBW: *Experimental Pulse NMR. A Nuts and Bolts Approach*. Addison Wesley, Massachusetts, 1981.

17. Fullerton GD: Basic concepts for nuclear magnetic resonance imaging. *Magnetic Resonance Imaging* 1: 39-53, 1982.

18. Fullerton GD: Physiologic basis of magnetic relaxation. In: Stark DD, Bradley WG (eds). *Magnetic Resonance Imaging*: 36-55. C.V. Mosby, St. Louis, 1988.

19. Fullerton GD and Cameron IL: Relaxation of Biological Tissues. In: Wehrli FW, Shaw D, Kneeland JB (eds). *Biomedical Magnetic Resonance Imaging*: 115-155. VCH Publishers, New York, 1988.

20. Gomori JM, Grossman RI, Goldberg HI, Zimmerman RA, Bilaniuk LT: Intracranial hematomas: Imaging by high-field MR. *Radiology* 157:87-93, 1985.

21. Haacke EM, Bearden FH, Clayton JR, Linga NR: Reduction of MR imaging time by hybrid fast scan technique. *Radiology* 158: 521-529, 1986.

22. Haacke EM, Bellon EM: Artifacts. In: Stark DD, Bradley WG (eds). *Magnetic Resonance Imaging*: 138-160. C.V. Mosby, St. Louis, 1988.

23. Hart HR, Bottomley PA, Edelstein WA, et al: Nuclear magnetic resonance imaging: Contrast-to-noise ratio as a function of strength of magnetic field. *AJR* 141: 1195-1201, 1983.

24. Heiken JP, Glazer HS, Lee JKT, et al: *Manual of Clinical Magnetic Resonance Imaging*. Raven Press, New York, 1986.

25. Hendrick RE: Image contrast and noise. In: Stark DD, Bradley WG (eds). *Magnetic Resonance Imaging*: 66-83. C.V. Mosby, St. Louis, 1988.

26. Henkelman RM, Bronskill MJ: Artifacts in magnetic resonance. *Reviews of Magnetic Resonance in Medicine* 2: 1-126, 1987.

27. Hennig J, Nauerth A, Friedburg H: RARE imaging: A fast imaging method for clinical MR. *Magn Reson Med* 3: 823-833, 1986.

28. Herfkens RJ, Higgins CB, Hricak H, et al: Nuclear magnetic resonance imaging of the cardiovascular system: Normal and pathological findings. *Radiology* 147: 749-759, 1983.

29. Hopkins AL, Yeung HN, Bratton CB: Multiple field strength in vivo T1 and T2 for cerebrospinal protons: A step towards intra-cerebral CSF pO2 and protein estimations. In: *Annual Meeting of the Society of Magnetic Resonance in Medicine*, 4th, Aug. 19-23, London, 1985.

30. Johnson GA, Glover GH, Karis JP, Shimakawa A, Herfkens RJ: Ultrahigh-resolution MR imaging with gradient refocused three-dimensional spin-warp sequence (Abstr). *Radiology* 161(P): 254, 1986.

31. Jones JP, Partain CL, Mitchell MR, et al: Principles of magnetic resonance. In: Kressel HY (ed). *Magnetic Resonance Annual 1985*: 71-111. Raven Press, New York, 1985.

32. Keller PJ: *Basic principles of magnetic resonance imaging*. General Electric Company, Milwaukee, 1988.

33. Kanal E, Wehrli FW: Signal-to-noise ratio, resolution, and contrast. In: Wehrli FW, Shaw D, Kneeland JB (eds). *Biomedical Magnetic Resonance Imaging*: 47-114. VCH Publishers, New York, 1988.

34. Kelly WM: Image artifacts and technical limitations. In: Brant-Zawadzki M, Norman D (eds). *Magnetic resonance Imaging of the Central Nervous System*: 43-82. Raven Press, New York, 1987.

35. Kneeland JB: Instrumentation. In: Stark DD, Bradley WG (eds). *Magnetic Resonance Imaging*: 56-65. C.V. Mosby, St. Louis, 1988.

36. Lauterbur PC: Image formation by induced local interactions: Examples employing nuclear magnetic resonance. *Nature* 242: 190-191, 1973.

37. Lee JKT, Dixon WT, Ling D, Levitt RG, Murphy WA: Fatty infiltration of the liver: Demonstration by proton spectroscopic imaging. *Radiology* 153: 195-201, 1984.

38. Lee JKT, Heiken JP, Dixon WT: Detection of hepatic metastases by proton spectroscopic imaging. *Radiology* 156: 429-433, 1985.

39. Macovski A: Volumetric NMR imaging with time-varying gradients. *J Magn Res Med* 2: 29-40, 1985.

40. Mitchell DG, Burk DL, Vinitske S, Rifkin MD: The biophysical basis of tissue contrast in extracranial MR imaging. *AJR* 149: 831-837, 1987.

41. McNamara MT: Paramagnetic contrast media for magnetic resonance imaging of the central nervous system. In: Brant-Zawadzki M, Norman D (eds). *Magnetic Resonance Imaging of the Central Nervous System*: 97-105. Raven Press, New York, 1987.

42. Ordidge RJ, Mansfield P, Doyle M, Coupland RE: Real time movie images by NMR. *Br J Radiol* 55: 729-733, 1982.

43. Packer KJ: The effects of diffusion through locally inhomogeneous magnetic fields on transverse nuclear spin relaxation in heterogeneous systems: Proton transverse relaxation in striated muscle tissue. *J Magn Reson* 9: 438-443, 1973.

44. Partain CL, James AE, Rollo FD, Price RR: *Nuclear Magnetic Resonance (NMR) Imaging*. W.B. Saunders, Philadelphia, 1983.

45. Pattany PM, Phillips JJ, Lee CC, et al: Motion artifact suppression technique (MAST) for MR imaging. *J Comput Assist Tomogr* 11: 369-377, 1987.

46. Purcell EM, Torrey HC, Pound RV: Resonance absorption by nuclear magnetic moments in a solid. *Phys Rev* 69: 37-83, 1946.

47. Runge VM, Clanton JA, Lukehart CM, et. al: Paramagnetic agents for contrast-enhanced NMR imaging: a review. *AJR* 141: 1209-1215, 1983.

48. Saine S, Frankel RB, Stark DD, Ferrucci JT: Magnetism: A primer and review. *AJR* 150: 735-743, 1988.

49. Sebok D, Pavlicek W, Weissman J, Weinstein MA: MR subtraction angiography: A nongated projection technique (Abstr). *Radiology* 161(P): 135, 1986.

50. Shaw D: The fundamental principles of nuclear magnetic resonance. In: Wehrli FW, Shaw D, Kneeland JB (eds). *Biomedical Magnetic Resonance Imaging*: 1-46. VCH Publishers, New York, 1988.

51. Sprawls P: Spatial Characteristics of the MR image. In: Stark DD, Bradley WG (eds). *Magnetic Resonance Imaging*: 24-35. C.V. Mosby, St. Louis, 1988.

52. Stark DD: The liver, pancreas, and spleen. In: Higgens CB, Hricak H. *Magnetic Resonance Imaging of the Body*: 347-372. Raven Press, New York, 1987.

53. van As H, Schaafsma TJ: Flow in nuclear magnetic resonance imaging. In: Petersen SB, Muller RN, Rinck PA (eds). *An Introduction to Biomedical Nuclear Magnetic Resonance*: 68-96. George Thieme Verlag, Stuttgart, 1985.

54. Von Schulthess GK, Higgins CB: Blood flow imaging with MR: Spin-phase phenomena. *Radiology* 157: 687-695, 1985.

55. Waluch V, Bradley WG: NMR even echo rephasing in slow laminar flow. J *Comput Assist Tomogr* 8: 594-598, 1984.

56. Wehrli FW: *Introduction to fast-scan magnetic resonance*. General Electric Company, Milwaukee, 1986.

57. Wehrli FW: Fast-scan imaging: principles and contrast phenomenonology. In: Higgens CB, Hricak H. *Magnetic Resonance Imaging of the Body*: 23-38. Raven Press, New York, 1987.

58. Wehrli FW: Principles of magnetic resonance. In: Stark DD, Bradley WG (eds). *Magnetic Resonance Imaging*: 3-23. C.V. Mosby, St. Louis, 1988.

59. Wehrli FW: *Advanced MR imaging techniques*. General Electric Company, Milwaukee, 1988.

60. Wehrli FW, Bradley WG: Magnetic resonance flow phenomena and flowimaging. In: Wehrli FW, Shaw D, Kneeland JB (eds). *Biomedical Magnetic Resonance Imaging*: 1-46. VCH Publishers, New York, 1988.

61. Wehrli FW, Perkins TG, Shimakawa A, Roberts F: Chemical shift-induced amplitude modulations in images obtained with gradient refocusing. *Magnetic Resonance Imaging* 5: 157-158, 1987.

62. Wesbey GE: Magnetopharmaceuticals. In: Wehrli FW, Shaw D, Kneeland JB (eds). *Biomedical Magnetic Resonance Imaging*: 157-188. VCH Publishers, New York, 1988.

63. Wolf GL, Burnett KR, Goldstein EJ, Joseph PM: Contrast agents for magnetic resonance imaging. In: Kressel HY (ed). *Magnetic Resonance Annual* 1985: 231-266. Raven Press, New York, 1985.

64. Young IR. Special pulse sequences and techniques. In: Stark DD, Bradley WG (eds). *Magnetic Resonance Imaging*: 84-107. C.V. Mosby, St. Louis, 1988.

Note that numerals in **boldface** indicate pages on which the term is defined.